INDIA'S WATERS

ADVANCES IN DEVELOPMENT AND MANAGEMENT

INDIA'S WATERS

ADVANCES IN DEVELOPMENT AND MANAGEMENT

MAHESH CHANDRA CHATURVEDI

CRC Press
Taylor & Francis Group
Boca Raton London New York

CRC Press is an imprint of the
Taylor & Francis Group, an **informa** business

CRC Press
Taylor & Francis Group
6000 Broken Sound Parkway NW, Suite 300
Boca Raton, FL 33487-2742

First issued in paperback 2017

© 2012 by Taylor & Francis Group, LLC
CRC Press is an imprint of Taylor & Francis Group, an Informa business

No claim to original U.S. Government works

ISBN-13: 978-1-4398-7466-0 (hbk)
ISBN-13: 978-1-138-11483-8 (pbk)

Library of Congress Cataloging-in-Publication Data

Chaturvedi, M. C.
    India's waters. Advances in development and management / Mahesh Chandra Chaturvedi.
        p. cm.
    Includes bibliographical references and index.
    ISBN 978-1-4398-7466-0 (hardback)
    1. Water resources development--India. 2. Water-supply--India. 3. Environmental management--India. 4. River engineering--Government policy--India. I. Title.

TC103.C439 2011
333.7900954--dc23   2011041493

Visit the Taylor & Francis Web site at
http://www.taylorandfrancis.com

and the CRC Press Web site at
http://www.crcpress.com

Contents

List of Figures

List of Tables

Preface

Water is crucial for life, environment, and economy. Development and management are needed in view of the mismatch between natural availability and demand. This is particularly important for countries of South and East Asia in view of their climatic–hydrologic characteristics.

As brought out in the accompanying first study, *India's Waters: Environment, Economy, and Development*, water resources development has been undertaken in India from time immemorial. One of the world's largest infrastructures for management of water has been developed.

Studies by the author have shown that, taking into account the characteristic Indian hydrological and physiographic characteristics, the policy and technologies used currently to develop India's waters can be revolutionized. The water availability can be almost doubled, and the hydroelectric potential can be increased several times and that, too, on a unique interspatial–intertemporal pumped storage basis. The current study brings out the proposed revolution in concepts, policies, technologies, and management of India's waters. The study has been kept brief to emphasize the proposed revolutions. A detailed study of India's waters has also been undertaken, but it is being kept pending, so that the emphasis on the proposed revolutions is not diluted.

The Ganga–Brahmaputra–Meghna (GBM) system is the heart of India's waters. A detailed study of the GBM system has also been undertaken and constitutes the third study of these three interrelated studies. It brings out the proposed revolutions in detail.

The proposals are not mere academic studies. Attempts are being made to implement the proposed revolutionizing of the water resources of India. Presentations of the proposed advances have been made to the Indian National Academy of Sciences and to the Central Water Commission, Government of India. The author is working closely with the Government of India on their implementation.

In view of the proposed revolutionary advances, it may be in order to give the author's background. This is presented in Appendix 2.

The author had the privilege of learning from, and later working with, some of the leaders of engineering and science in India and internationally—Dr. A.C. Mitra, Dr. A.N. Khosla, Dr. K.L. Rao, Prof. Hunter Rouse, and Prof. Roger Revelle. This is gratefully acknowledged and the book is dedicated to his gurus.

Acknowledgment is due to a large number of professionals and doctoral scholars, with whom the author worked. As with the other two books, the present book owes much to the support of a number of younger colleagues: S.K. Pathak, B.S. Mathur, K.N. Duggal, V.K. Srivastava, D.K. Srivastava,

R. Devi, S.D. Khepar, Y.C. Arya, R. Singh, R.K. Prasad, U.C. Chaube, M.K. Munshi, B.N. Asthana, D.K. Gupta, (late) P.K. Bhatia, L. Singh, A.V. Chaturvedi, E.A.S. Sarma, P. Deb, and C. Thangraj, to name a few, who are former doctoral scholars and later senior academicians or real-life engineers who contributed much to author's education. The author owes much to his colleagues from Harvard University, (late) Prof. J.J. Huntington, Prof. Peter Rogers, and Prof. J. Briscoe.

The Ford Foundation, Delhi, provided generous support to the author in modernizing water resources development in India. Dr. Sadik Toksaz and Dr. R. Lenton, in charge of the Ford Foundation water program, New Delhi, were very supportive of this activity, with Dr. Lenton even taking classes at the Indian Institutes of Technology (IIT) Delhi. Acknowledgment is gratefully made of their support. Acknowledgment is also due to all the scholars whose work is referred to in the book as the author's understanding of the subject owes much to them.

The support by Prof. B.N. Asthana, a student, professional colleague, and collaborating scholar, is also acknowledged. He kindly undertook to review the entire work and minimize the errors. The study would not have seen the light of the day without his support. Shri D.C. Thakur took over the responsibility of editing the drawings and undertaking the publication process with the publishers, which is acknowledged and deeply appreciated. Acknowledgment is also made of the support of Chairman, Central Water Commission, and his colleagues, who were kind enough to give the necessary information regarding the subject.

The study, along with the accompanying works, was undertaken in India and at the Harvard University, where the author spends his summers to wrap up his yearly work in India. Thanks are due to Prof. Peter Rogers and Dean V. Narayanamurti for the long and continuing Harvard sojourn.

Finally, acknowledgment is gratefully made to the author's wife, (late) Prof. Vipula Chaturvedi, who had constantly encouraged the author in the writing of these books, for collaborating on several papers and a book, and for bearing with him for the neglect that such an undertaking entails of personal relations. The recent loss of the author's wife has been a most grievous blow to him.

Mahesh C. Chaturvedi
New Delhi/Cambridge, MA

1

Introduction

Water is an important component of the environment. It is crucial for life, economy, and sustainable environment itself. The earliest human settlements took place in regions of benign climate and water availability such as Mesopotamia, Egypt, India, and China. The largest settlements gradually grew in China and India, which offered plenty of land and water, and they emerged as the regions of greatest development until about the mid-eighteenth century. However, some settlements have lost their water independence, and they are currently among the world's most underdeveloped regions with the highest concentration of population.

A new world revolution has started taking place. Both these countries, China and India, are gradually growing, with high rates of economic development and population growth. There are many challenges. One is the management of the environment, in which management of water is crucial to these countries on account of their environmental characteristics. With the high and increasing population, and economic activities, they will be experiencing a very high level of strain regarding water. Focusing on India, which is our subject of study, the scene becomes alarming. India sustains about 20% of the world's population, but accounts for only 2.4% of the world's surface and 3.5% of the freshwater resources. There is no substitute for water.

Development of water resources is important in any society, but is even more important in India, as in other South Asian countries and China, in view of the arid monsoon climate and agrarian economy. The agriculture demand varies, depending on the climatic characteristics, particularly the rains. For example, little effort has to be made for supplying water to crops above the natural supplies of rains in Western Europe, whereas in India, on account of the climatic conditions, irrigation has to be provided to agriculture, which has the dominant water demand, about 80%, compared to about 10% in Western Europe and the United States, in view of their climatic conditions. With the present commitment of about 50% of the utilizable water resources in India, currently estimated at 1086 km^3, it has only been possible to provide irrigation to only about 37% of the sown area. The quality of water service is extremely poor, leading to one of the poorest yields, even among the developing countries. In a recent study of yields in 93 developing countries, excluding China, the average yield was 1951 kg/ha. The yields in India were only 1600 kg/ha; whole yields in China, with about the same irrigation intensity and cultivated land area, were about 4329 kg/ha. The yields in industrialized countries are around 6000 kg/ha (World Resources Institute

1996). This not only leads to poor economic returns but also imposes a severe burden on the environment.

The scene is bad in all sectors of water use. The most important area of water demand and supply is drinking water. Potable drinking water is not available even in metropolitan cities, much less in rural areas or villages. A U.S. scientist reports the scene as follows. "A friend of mine lives in a middle-class neighborhood of New Delhi, one of the richest cities in India. Although the area gets a fair amount of rain every year, he wakes in the morning to the blare of a megaphone announcing that the fresh water will be available only for the next hour. He rushes to fill the bathtub and other receptacles to last the day ... My son, who lives in arid Phoenix, arises to the low, schussing sounds of sprinklers watering verdant suburban lawns and golf courses. Although Phoenix sits amid the Samoran Desert, he enjoys a virtually unlimited water supply" (Rogers 2008). The "friend" happens to be the author, with whom Peter Rogers, a Professor at Harvard University, studied the water resources of India at one time (Chaturvedi and Rogers 1985). What Rogers did not mention is that the "friend" then proceeds to boil the water to meet the drinking water supplies because the public water supply is not potable. In addition, often, the announcement is "Aj pani nahin aega" (which figuratively means "Water will not come today"). The pitiable condition of the people who are not as well-off or who live in the villages can be easily imagined.

Even in this process, all the river waters have been diverted to meet the irrigation supplies, and with the growing urbanization and even with only a limited industrialization as yet, the rivers have been turned into fetid sewers over large stretches. According to latest official estimates, "available supplies on certain premises will be matched, if not exceeded by the demand by the year 2050. Water stress-conditions will be experienced in many parts of the country unless remedial measures are taken in time" (NCIWRD 1999, p. 15). In addition, this is only the beginning of the story. The future perspective is extremely disconcerting.

There is another serious and distinguishing challenge. The water availability characteristics in Western Europe, where scientific advances and industrialization started taking place, and in India or other countries of Southeast and East Asia, such as China, are markedly different. The latter have monsoon climate. Whereas availability of water varies over the year everywhere, in monsoon climates this is highly intensified. It is further aggravated in India by the hot weather conditions. A long dry period would suddenly experience severe monsoon rains, leading to most severe floods. Even in the rainy season, the rains are concentrated in a few short periods. About 80%–90% of rains are concentrated in 45 days, and that, too, over a few days. The scene is further compounded on account of the fact that there are marked spatial and temporal variations. For instance, at one end of India, Cherrapunji is the wettest place on earth. On the other hand, in the same river basin, Rajasthan is a desert. Even in Cherrapunji, people may be washed away by abundant water during the monsoons, while they have to collect adequate drinking water, much less water for crops, during nonmonsoon periods. Furthermore, the availability is very uncertain. Annual variations in most parts of the country can easily be ±30% of the mean. There is even greater monthly variation.

Another serious predicament may be added—the climate change. It has been estimated by renowned climatologists that the rainfall, besides displaying considerable spatial and temporal changes, will be diminished by about 60% in Northern India in about a century (Cline 2007). The disappearance of glaciers in the Himalayas is another frightening scenario. Unfortunately, not much attention has been paid to these ominous warnings so far.

As observed by the author, a long time back, "the serious predicament that awaits us can be avoided only if there is a revolution in our concepts, organization, capabilities, and immediate action is taken to meet the critical situation ahead" (Chaturvedi 1976). Unfortunately, the current official perspective, as represented by the Official Commission (NCIWRD 1999), to give one example, fails to get out of the trap of the current concepts and poor planning of the water resources of India and address the resolution of the serious challenge. This is equally true of some recent studies relating to India's waters (Kumar 2010; Pasucal et al. 2010). They only present India's waters in terms of current concepts, technology, and institutions.

Therefore, there is an urgent need to study the subject creatively. Our objective, based on a long professional and academic experience, is to contribute to revolutionizing the subject.[1] This is the objective of this and the companion studies (Chaturvedi 2011b,c,d).

A brief review of the current concepts and attempts of development and the needed revolution is undertaken in the book, attempting to inform the people, the concerned scientists, and decision makers. Detailed study of one major river basin, Ganga–Brahmaputra–Meghna (GBM), is undertaken in an accompanying study, which also gives the modern conceptual advances leading to societal environmental systems management, as proposed by the author, with some novel technologies, which revolutionizes the development of water (Chaturvedi 2011a,c,d). Therefore, we proceed as follows. A brief introduction is given to the environmental conditions of India, including the climate change, to bring out the dynamics of the environment and future challenges. Next is an overview of the development of the water resources of India, followed by current official perspective of future challenges and proposed responses. In contrast, a scientific and creative approach is proposed. The possibility of some novel technologies, taking into account India's physiographic and hydrologic characteristics, is brought out. We demonstrate that it is possible to revolutionize the development and management of water through some novel technologies, enabling us to meet the formidable challenges. Modern scientific advances in the field of water resources planning and management are also presented briefly to bring out the revolution in this respect also. The emphasis, essentially, is on the development of the water resources. The other components of water cycle—use and management of the return flows—have not been emphasized. This does not mean that they are of any less significance.

However, technology is only one facet of activity. The economic activities have to be conjunctively advanced, no doubt spurred by technology. Even more importantly, the governance and culture have to be transformed. Development of water to meet the needs of life and economy is extremely important in the context of the urgent need to revolutionize India. Technologies and policies are proposed to be radically changed. However, the central issue is action. That is the objective of the three studies by the author on the subject. Attempts for implementation of the proposed suggestions are also being made. We are closely engaged with the Government of India and reputed Indian and U.S. engineering institutions on the implementation of the ideas presented in these studies.[2]

Much of the writing on water resources is undertaken by the Western scientists.[3] A basic perspective of development, in their view, and in the development of water is eradication of poverty. While that is an extremely important and priority issue for us also, our perspective of development is entirely different. For us, development means attainment for India, representing about 20% of the world population, its due state of equality. Extended logically, it means global leadership in concepts and technology, in conjunction with the rest of humanity. However, as Sen (2009) notes, action is urgently needed to eradicate our current shameful state. This concept pervades all our statements on water resources development.

Notes

1. The author has designed almost all the major dams in the Ganga basin or has been on their Board of Consultants since 1946.

2. A project has been formulated by the author, under which the five Indian Institutes of Technology (IITs) of the GBM basin are proposed to undertake detailed studies of the basin. This has been approved in principle by the Ministry of Water Resources, GOI, and is in the process of implementation. Similar action was proposed to the Honorable Minister of Environment and Forests, GOI. He has implemented the suggestion and an agreement has been reached between the IITs and the Ministry of Environment and Forests, GOI, for the preparation of the Ganga River Basin Master Plan on July 6, 2010.

   The University of Iowa, Iowa City, IA, the author's alma mater, conducts a course entitled "International Perspectives in Water Resources Science and Management." They will be studying the

GBM basin for a fortnight from Dec. 27, 2011 to Jan. 15, 2012, organized by the IITs of the Basin, in collaboration with the author.

3. Reference to one of the latest contributions may be made: Rogers, P. (ed.). 2011. There is not a single contributor who is from the Third World, although the focus is their challenge.

2

Environmental Characteristics

2.1 Introduction

Each country has well-defined environmental characteristics. It is important to take them into consideration because they define the potential and challenge of development of water-related resources. Surprisingly, some unique Indian environmental characteristics, which lead to some revolutionary possibilities of development of water, have been totally ignored. There is another feature that should be emphasized. As climate change has demonstrated, climate is not a static state. The issue of climate change has to be considered along with the societal challenge, as action has to be taken urgently to face this challenge as well.

2.2 Geographical Characteristics

India, that is, Bharat (8°4′N–37°6′N and 68°7′E–97°25′E), embraces a major part of the South Asian realm. Girdled by the young fold mountain chain on its NW, N, and NE and washed by the Indian Ocean and its two main arms, the Arabian Sea and the Bay of Bengal on the SW, S, and SE, India is a well-defined geographical unit though with contrasting features.

The Republic of India has a land frontier of 15,200 km and a coastline of 5700 km. With its N–S and E–W extent of 3200 and 3000 km, respectively, and with an area covering about 3,270,500 km², it ranks seventh among the countries of the world in geographical area. The location and characteristics—a large peninsula with high mountains on the North—create a unique hydrologic-climatic environment.

2.3 India—Geological History

Geologically, the country represents a monumental assemblage of land patterns, varying in age from Pre-Cambrian to the recent. The Peninsular massif is the core, around and on which different acts of geological drama were staged, and all have left their imprints in some form or another. This massif, a part of the supercontinent "Gondwanaland" until its rupture and drifting sometime in the mid-Mesozoic era (about 200 million years ago), lays somewhere near the South Pole along with Australia, South Africa, and South America.

Broadly speaking, six major geological phases have been considered responsible for the formation of the Indian subcontinent. The first phase is marked by the cooling and solidification of the upper crust of the Earth's surface in the Pre-Cambrian era (before 600 million years). The fifth phase, in the mid-Mesozoic era (200 million years ago), is recognized as the world's major event, which comes in the form of fracturing and drifting of the continental mass of "Gondwanaland," essentially defining the current scene. The close of the Mesozoic (70 million years ago) witnessed one of the greatest volcanic eruptions, the Deccan lava flows, covering more than 500,000 km^2 area. This was followed by the first phase of the tertiary progeny—the Karakoram phase, which was characterized by the coming together of the two land masses, the Angaraland in the north and the Peninsular India in the south, under the oscillatory movement of the continental masses to and from the pole.

The movements responsible for the three parallel folds of the Himalayan Proper took place subsequently during the Oligocene (25–40 million years ago), mid-Miocene (1 million years ago), and Post-Pliocene (750,000 years ago). Contemporaneous had been the formation and alluviation of the Tethys geosynclines or the Indo-Ganga trough and the tectonic troughs in the Himalaya, especially the Kashmir valley. Alluvial and glaciofluvial deposits (Pliocene–Holocene) filled up these troughs. Considerable drainage derangements must have taken place before the present system could have established. Mention must be made of the Rajmahal–Garo gap or the Malda Gap (Pleistocene) as well as of the later upheaval in the Indo-Ganga divide, which dismembered and diverted the whole of the drainage of India, east of Aravalli, to the Bay of Bengal, originally going NW through the Indo-Brahm or Siwalik River. Other major derangements may be noted, such as the Narmada–Tapi troughs and the foundering of the west coast.

The buried extent of the peninsular massif has imparted a semicircular trend to the young fold mountains, with syntaxial bends around the Salt range in the NW and Namcha Barwa in the NE. The channel of Ganga is also in synchrony with the shape of the peninsular foreland and has perhaps reached its southward shift, washing at places the old bedrocks.

2.4 Physiography

Complex land-building forces and erosion processes have been at work in India throughout its geological past, and the face of the country is as complex as its making. For simplicity, based on stratigraphic and tectonic history and relief along with the erosion processes, four macro physiographic regions have been distinguished: (1) the Northern mountains, (2) the Great Plains, (3) the peninsular uplands, and (4) the Indian coasts and islands. Sometimes, further detailing into seven divisions and 20 subdivisions, as shown in Figure 2.1, has been done.

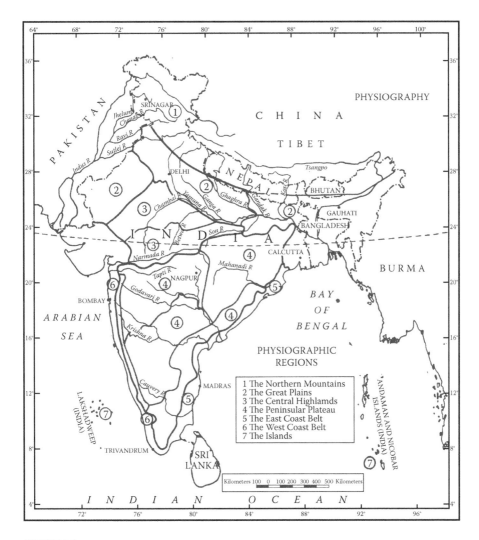

FIGURE 2.1
India physiographic map. (From Chaturvedi, M.C., *Water—Second India Studies*, Macmillan, New Delhi, 1976.)

2.4.1 The Northern Mountains

The region extends all along the northern border of the country, from the eastern border of Pakistan to the frontiers of Burma, for about 2500 km, with an average width of about 240 km. Occupied by Himalayan ranges and its offshoots, it covers an area of about 500,000 km². Three major fold axes represent the Himadri (Greater Himalaya), Himanchal (Lesser Himalaya), and the Siwaliks (Outer Himalaya), extending almost uninterrupted along the entire length. Mighty but older streams such as the Indus, Sutlej, Ganga, Kali, Kosi, and the Brahmaputra have cut through steep gorges to escape into the Great Plains and have established their antecedents. It is impressive that three of the world's largest rivers, Indus, Ganga, and Brahmaputra, originate nearly from about the same place and flow east, south, and west, cutting through the Himalayas to the plains. The troughs intervening in the ranges are occupied by the longitudinal valleys of the streams in their upper reaches. The main unifying factor is the parallelism of the three axes extending east–west. Himadri, the asymmetrical and the northernmost range of the Himalaya, owes its scenic beauty to glaciers and lofty snowy peaks. This range has a granite core, flanked by metamorphosed sediments. It is credited for having the world's 14 highest peaks, ranging between Jano (7710 m) and Everest (8848 m).

The Himachal forms the central chain composed mainly of highly compressed and altered rocks varying from Algonkian or Pre-Cambrian to Eocene in age. In general, the alternate ranges and valleys acquire an elevation of about 5000–1000 m, respectively. Its asymmetrical structure at places provides it more or less hogback look. It differs from Himadri in its more regular and lower elevation.

The Siwaliks represent the outermost range of the system, with roughly a hogback appearance, a steeply sloping southern face and a gently sloping northern face. These newer and river-borne deposits deriving from the rising Himalaya represent the most recent phase of the Himalayan orogeny, that is, from Middle Miocene to the Lower Pleistocene. The range, bordered on the north by flat-floored structural-longitudinal or erosional valleys called the *Duns*, is characterized by fault scraps, anticlinal valleys, and synclinal ranges. Apart from these longitudinal subdivisions, the Himalayas exhibit regional characteristics.

2.4.2 The Great Plains

This aggradational plain covers about 700,000 km² of the surface area, with the Ganga and the Brahmaputra forming the main drainage axes in the major portion. The thickness in the alluvial sediments varies considerably with the maximum in the Ganga plain. The variation in thickness largely depends on the alluvial-morphological processes. The cones of the Kosi in the north and the Son in the south exhibit greater alluvial thickness, whereas the intercone areas have relatively shallower depths. The physiographic scenery varies from the extremely arid and semi-arid landscape of the Rajasthan Plain to the humid and per-humid landscape of the delta and the Assam valley in the east. The Delhi ridge is a subdued extension of the Aravallis. Topographic uniformity, except in the arid western Rajasthan, is a common feature throughout, although the nature of the materials brought down by the rivers varies significantly, resulting in the local geomorphologic variations. The Brahmaputra, the Ganga, and the intervening rivers carry more sand than silt, have formed long levees, and have also raised their beds. Even the partial washing away of these levees during the high floods submerges the extensive low-lying plains, causing immense damage to life and property. With an average elevation of about 150 m, ranging from almost nothing (Bengal Delta) to nearly 300 m (Punjab and Upper Ganga Plains) near the foothills, the area is characterized by extremely low gradients.

Along the northern margin of the plain lie two narrow but distinct strips—the Bhabar and the Tarai. The Bhabar (a piedmont plain: 10–15 km wide) is composed of unassorted debris from the Himalayas. The surface streams disappear in this zone of boulders and sands. Immediately below the Bhabar is the 15–30-km-wide, relatively low-lying, Tarai region, which is characterized by finer sediments, a natural forest cover, emergent and ill-defined water channels, low gradients, and a high water table (ranging from a few meters to about 5 m below the ground), resulting in swamps and marshes.

The *Gangodh* (i.e., the Ganga alluvium) is distinguishable in two types, that is, the *Khadar*, the strip covered with recent alluvium and liable to frequent inundation and siltation, and the *Bhangar*, comprising older alluvium seldom prone to inundation. Changing river courses in the area of frequent overflooding present interesting geomorphic processes in the plains. The southern margin of the plains, being in contact with the southern uplands, is often encroached by the projections of the peninsular mass, sometimes up to the bank of the Ganga.

2.4.3 The Peninsular Uplands

This morphologically polygenetic, complex, and relatively stable landmass extends from the southern margin of the Great Plains onto coastal margins of the country and covers an area of 1.6 million km². It presents a natural landscape of detached hills, summit plains, entrenched narrow as well as aggradational wide valleys, and series of plateaus, peneplains, and residual blocks. One of the major physiographic elements, the SW–NE Aravalli hills, is a relic of the world's oldest fold mountains. Considerably dissected and almost detached by the Banas and the Luni in the central parts, it spreads like a fan toward the north, sending projections up to Alwar and adjoining parts and into the Udaipur region to the South, which has the highest elevation at Mt. Abu (about 1772 m). The Vindhya–Satpura alignment, owing to a steep scarp and a range-like characteristic, to the Narmada trough extends east–west from the Sahyadri in the west to the Maikal in the east. Most of the Satpura–Vindhya region is overlain by Deccan Trap in the west, with a general horizontal disposition. The Vindhyas show a somewhat folded structure, particularly in the western section. The gentle gradient of the Vindhyas in the north and the steep step-like face overlooking the trough to the south are another set of distinguishing features, whereas the Satpuras possess steep gradients toward the Narmada valley in the north and Tapti (Tapi) valley in the south. Average elevation rises to about 300 m, with occasional detached summits rising to over 1000 m (Panchmarhi and Amarkantak). Hemmed, in between the Aravallis and the Vindhyas, is the triangular, dissected sedimentary surface of the Vindhyan basin that has the imprints of crisscross faults and joints. The Vindhyan range largely marks the watershed between rivers into the Great Plains and the other streams flowing toward the south or west. The exception to it is the Son River, which is an accidental consequent stream.

The Sahyadri was probably a central water divide of an older extensive landmass. With its north–south trend, it serves even today as a divide between the Bay of Bengal and the Arabian Sea drainage. The Thalghat, Bhorghat, and Palghat are three major gaps in this alignment that, since early times, have served as a negotiating link between the coastal lands and the rugged plateau country. With a steep wall-like appearance overlooking the west coast, they send out projections (Balaghat, Mahadeo range, etc.) to the eastern plateau country, with relatively minor breaks in the slope. Its maximum elevation is obtained at and south of its junction with the Eastern Ghats in the Nilgiri knot (2636 m) and the Animalai–Palni hills (2695 m), respectively.

On the eastern margin of the peninsular uplands lie much more discontinuous but similar ranges from Mayurbhanj (Orissa) to the Charnockite hills of the Nilgiris. Almost all the major streams such as the Godavari, the Krishna, and the Kaveri, taking their rise from the Western Ghats or the Sahyadri, have cut extensively through the Eastern Ghats to escape into the Bay of Bengal.

Encompassed between these ranges are numerous gorges, waterfalls, wide alluvial valleys (Wardha–Wainganga plain) and structural-cum-erosional basins (Chhattisgarh and Cuddapah); insignificant aerially but of great economic importance are the Gondwana troughs or basins, containing over 98% of the coal reserves of the country.

To sum up, the story of the peninsular landscape consists of several cycles of denudation, sedimentation, and igneous activities in harmony with orogeny, epeirogeny and cymatogeny, effusion, metamorphism and deep-seated rocks, tearing, eustatism, and widespread resurrection.

2.4.4 Indian Coasts and Islands

The Indian coasts vary widely in their structural and surface characteristics. The west coast is much narrower, except around the Gulf of Cambay and the Gulf of Kachchh, where, partly as a result of sedimentation and partly on account of the isostatic adjustments, the plains are wide enough. The Girnar Hills, the volcanic cones, appear to be an extension of the older landmass probably separated during the foundering of the west coast. The two gulfs might have been linked together. The silting of the link would have given rise to the Gujarat Plains. The silting left no scope for depositional action of the rivers of the west coast, and it still retains its narrow extent throughout is length south of Gujarat to the Cape Comorin. It is only in the extreme south that they are somewhat wider along the south Sahyadri. The backwaters are the characteristic features of this coast. The east coast plains, in contrast, are broader, associated with the depositional activities of the rivers, partly owing to the change in their base levels. Extensive deltas of the Mahanadi, the Godavari, the Krishna, the Kaveri, etc. are characteristic features of this coast. The progression of the deltaic plains into the sea is still continuing. Even tertiary gravels and sands are to be observed in this region. Physiographically, the coasts are subdivided into the following:

1. Gujarat Coast
2. West coast
3. East coast

There are two groups of Islands, the Arabian Sea Islands and the Bay of Bengal Islands. They are not discussed because the study is confined to the mainland.

2.5 Drainage System

The drainage system corresponds to the geological history (Figure 2.2). The Indian drainage may broadly be divided into the Bay of Bengal drainage and the Arabian Sea drainage, with distinct water partings lying approximately along the Sahyadri, Amarkantak, Aravallis, and the Sutlej–Yamuna divide. It is also distinguished as the Himalayan drainage and Peninsular drainage although several of the peninsular streams such as the Chambal, the Betwa, and the Son, much older in age and origin, form part of the Himalayan drainage system.

Over the geological period, Indus and Brahmaputra flew west and east, respectively, along the trough, whereas the Ganga and its tributaries, and the tributaries of Indus cut through the Himalayan ranges to flow south. Over geologic time, several climatic and hydrologic changes took place. The Himalayan mountain system was cut, heavily eroded, and the sediment load was deposited in the trough between the Himalayas and the Vindhyas, leading to the Great Indo-Ganga plain. These alluvial plains, an outcome of these

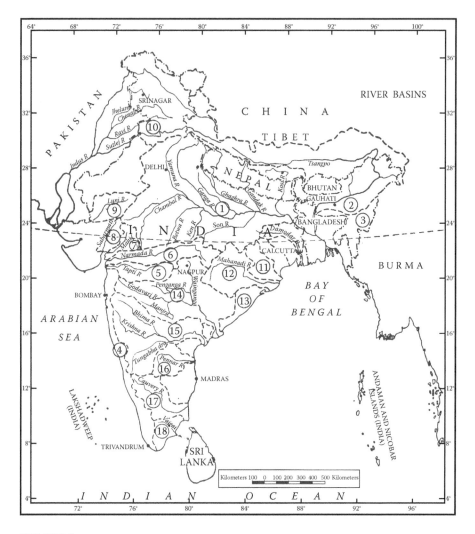

FIGURE 2.2
River basins. (From Chaturvedi, M.C., *Water—Second India Studies*, Macmillan, New Delhi, 1976.)

hydrological processes and geological activities, are a very flat, rich, and fertile land consisting of unconsolidated river-borne sediments thousands of meters deep and rich in groundwater.

The upper Indus and its longest tributary, the Sutlej, rise within 130 km of each other yet envelop the entire Western Himalayas before they meet at Mithankot, well out in the plains and over 1000 km to the west. The Indus system drains into the Arabian Sea.

In a symmetrical fashion to the east, the Tsangpo–Brahmaputra and Ganga envelop virtually the entire Eastern Himalayas, and the Ganga collects the intervening drainage. Again, the Tsangpo and the Ganga rise within 130 km of each other and of the sources of the Indus and Sutlej.

To explain this remarkable pattern, attention is directed to the Kailash Range overlooking the twin lakes of Mansarovar and Rakas. Although Mount Kailash (678.44 m), which lies midway between the world's two highest mountains, Everest (894.1 m) and K2 in the Karakoram (870.1 m), is considerably lower than either of them, the Kailash range forms one of the great water divides of Asia. Northwest of this range, water flows 3200 km via the Indus to reach the Arabian Sea. Southeast, drainage via the Tsangpo–Brahmaputra leads over 2736 km to the Bay of Bengal. To the southwest, Lake Mansarovar and Rakas feed the Sutlej, whereas directly south of the lakes, rises Karnali, which drains into the Ganga. It is no wonder that Mount Kailash is sacred to Hindus and Buddhists, who regard it as Shiva's paradise. The Kailash range thus holds the key to the Himalayan drainage, and its elevation in the very recent geological past would explain the symmetry to the east and west.

A network of large perennial rivers, fed by melting snow from the Himalayas and smaller channels, serves the plains from the north. There is an almost imperceptible drainage between the Ganga and the Indus drainages of very recent origin. The Indo-Ganga plains are very rich in groundwater of the purest quality.

Although the three major rivers of the world—Indus, Ganga, and Brahmaputra—have a common geological history, their geophysical characteristics are significantly different. Indus and its long major Sutlej flow over a long distance in the Himalayas, until they debouch in the Indo-Ganga plains. On the other hand, in view of the geological history, Ganga and Brahmaputra, and their tributaries, have a very steep longitudinal section. This is shown in Figure 2.3 and has important implications, which have been neglected so far. We will develop some novel plans, as will be discussed later, on the basis of these characteristics.

The Peninsular region is geologically much older and is served by a number of rivers, taking off from the Western Ghats, flowing almost due east, almost in a parallel formation, cutting through the Eastern Ghats, and debouching in the Bay of Bengal. The main rivers are Godavari, Krishna, and Cauvery, from north to south in that order. A significant characteristic of the system is that two major rivers, Narmada and Tapi, north of the peninsular rivers and south of the Himalayan Rivers, flow in the Southern Highland region from east to west, and another major river, Mahanadi, dividing the southern peninsular region and the Ganga–Brahmaputra–Meghna plain in the north, flows east at about the same longitude.

One major southern tributary of the Ganga basin, Son, takes off in the Southern Highlands, close to Narmada, north of it and joins Ganga near Patna (historical Pataliputra), where two major tributaries of Ganga also join from the north, bringing out a major augmentation of the discharges of Ganga. Similarly, a major tributary of the Godavari, Wanaganga, flows

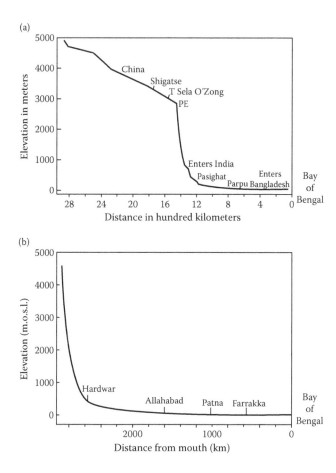

FIGURE 2.3
Longitudinal section of (a) Brahmaputra and (b) Ganga. (From Bruinzeel, 1989, pp. 4 and 5.)

down due south, taking off very close to Narmada on the south. This feature, again, has important implications, which have not been taken note of, and is utilized in our novel technologies.

In contrast to the Himalayan Rivers, the rivers of the peninsular region are not snow fed. The region is not as rich in groundwater. The return flows are, therefore, small and, for both these reasons, have very reduced summer flows and much variation in the monsoon and summer flows. Development of groundwater is also comparatively difficult and less economical. However, the geological formations present opportunities for excellent surface storages, small and big.

In view of the climatic conditions, agriculture can be pursued around the year, with two or three crops, depending on the availability of water. Thus, management of water is the key to the agricultural development and productivity.

2.6 Climate

The climatic characteristics are brought out in the pattern of temperature and annual pan evapotranspiration shown in Figures 2.4 and 2.5, respectively. Although a considerable portion of the country belongs to the subtropical zone, as a whole, it has the characteristics of a tropical monsoon climate, mainly because of the Himalayas, which function as an effective meteorological barrier. Rhythm is the keynote of the monsoon climate. The two

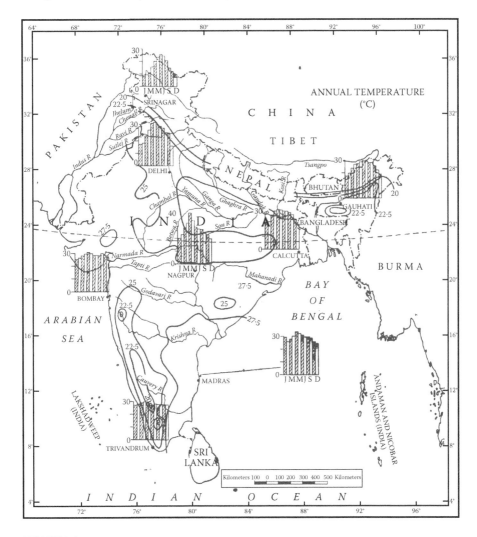

FIGURE 2.4

Annual temperature. (From Chaturvedi, M.C., *Water—Second India Studies*, Macmillan, New Delhi, 1976.)

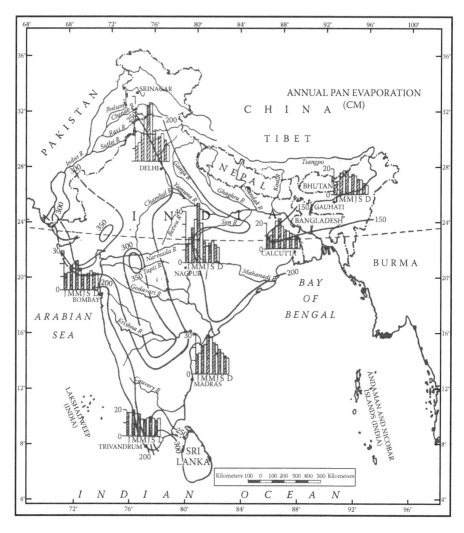

FIGURE 2.5
Annual pan evaporation. (From Chaturvedi, M.C., *Water—Second India Studies*, Macmillan, New Delhi, 1976.)

seasons, the summer and the winter, roughly correspond with the culmination of the sun from the Southern to Northern Hemisphere and vice versa, and their associated monsoon regimes. Hence, the Indian climate can be discussed under two heads—the summer and winter monsoons. The imbalance in their regime owes definitely to the differential heating and cooling as the temperature starts rising much before the vernal equinox, thus cutting short the winter season by about a month. Similarly, the retarded terrestrial radiation pushes its commencement further by about a month from the autumnal equinox, reducing the duration to a little over 4 months. The summer season is bifurcated as dry and wet or humid, roughly stretching from March to mid-June and mid-June to mid-October, respectively.

The winter season is characterized by lower temperatures, sometimes below freezing in some places, low humidity, and scanty rainfall. The following dry summer is marked by a sharp rise in temperature and a consequent decrease in relative humidity, giving rise to hot winds, locally called "Loo" over some parts of the country. It may be noted that this transition between the winter and summer season is essential, as it sets the stage for the outburst of the summer monsoon, which curbs the upward tendency of the temperature, sometime in late May and June. The wet summer, or the rainy season, is characterized by high humidity and fairly high temperature, creating unusual sultry conditions. A short transition is also experienced between the changes from the summer monsoon to winter, sometimes during October–November.

The summer monsoon is experienced as a "burst" and "break." This is explained in the withdrawal of the quasi-permanent westerly jet stream of the upper troposphere from above the northern part of the country to the North of Tibet by early June. Subsequently, an easterly jet stream develops in the lower stratosphere above the easterlies of the upper troposphere, generally lying over 15°N. An unstable trough of low pressure in the upper air extends from the Bay of Bengal around the Andaman Islands toward "heat low" of NW India. The irregularities in these processes are directly associated with the two vagaries, that is, the burst and break in the monsoon.

2.6.1 Summer Season and Summer Monsoon

The monsoons are controlled by the seasonal alternating low- and high-pressure conditions over the land and the sea primarily due to the differential terrestrial heating. The temperature registers a sudden upward change. The condition of uncertainty prevails throughout the country until the seasonal low is replaced in the NW by seasonal highs by the end of May, controlling the movement of air masses. The temperature occasionally fluctuates because of the pre-monsoon thundershowers, but the rise is checked only with the outburst of monsoons. Arial significance of pre-monsoon rainfall varies from region to region. For instance, it shares 26% of annual precipitation in Assam, 17% in Bengal, 15% in Mysore, 13% in SE Tamil Nadu, 4% in West UP, and 0.7% in Gujarat.

The spatial distribution of the average summer seasonal temperature ranges from 20°C over the high altitudes of Himalaya and south Sahyadri to more than 30°C along the desert frontiers and Tamil Nadu coast. The relative humidity rises to above 80%. The rainfall over the country is primarily orographic, associated with tropical depressions originating in the Bay of Bengal and the Arabian Sea. The summer monsoon accounts for most of the rainfall, with uneven spatial distribution almost in sympathy with the orography (Sahyadri, Eastern Himalaya, Meghalaya 200 cm and parts of Karnataka–Maharashtra Plateau, Punjab, Western UP, and Rajasthan less than 60 cm). Its uncertainty of occurrence marked by prolonged dry spells and fluctuations in seasonal and annual amount causes a serious problem indeed.

Peculiarly enough, the westward decreasing rainfall, in the northern part of the country, shows a reverse trend in its percentage share in the total annual precipitation, which varies from 66% in Assam to 96% in Gujarat. The only exception in the high share of summer monsoon rain in the country is the southeastern Tamil Nadu region (about 34%), where retreating monsoon has greater influence (about 39%).

The coefficient of variability reveals the increasing fluctuation in seasonal rainfall from east to west in the Great Plains and the Northern Peninsula (Sambalpur 14.6%, Jalpaiguri 16.3%, Silchar 17%, Nagpur 23.4%, Allahabad 28%, Delhi 33.3%, Bikaner 48.4%, and Bhuj 60%), whereas on the Southern Peninsula Uplands, it is from west to east (Kozhikode 16.3%, Bangalore 21.7%, and Madras 30.3%). Thus, the rainfall reliability, in general, is inverse to the total amount. The areas with marginal rainfall suffer worst.

2.6.2 Winter Season and Winter Monsoon

The winter monsoon commences after a short transitional span and lasts from mid-October to February. Associated with decrease in temperature and relative humidity, this period is responsible for a small amount (below 10% of annual total) of welcome rainfall in the country. In Tamil Nadu, Kerala, and Mysore, however, the winter monsoon rain is more significant (SE Tamil Nadu 39%, Madras coast 25%, Mysore 21%, and Malabar 16%). Northern India is affected by the western disturbances mainly between December and February, with decreasing intensity from west to east, and the share of precipitation varies accordingly (Kashmir 22%, Punjab 12%, Western UP 6%, Bengal 2%, etc.). These disturbances often cause snowfall in the Himalaya and subsequent sweep of the cold wave over the Great Plains, occasionally bringing down the temperature below freezing point.

2.7 Precipitation and Evapotranspiration Characteristics

The precipitation and evapotranspiration characteristics are important determinants of water resources development and, therefore, may be briefly brought out and illustrated. The annual isohyets are shown in Figure 2.6. The annual rainfall extends to 250 cm along the entire west coast and Western Ghats and over most of Assam and sub-Himalayan West Bengal. The annual

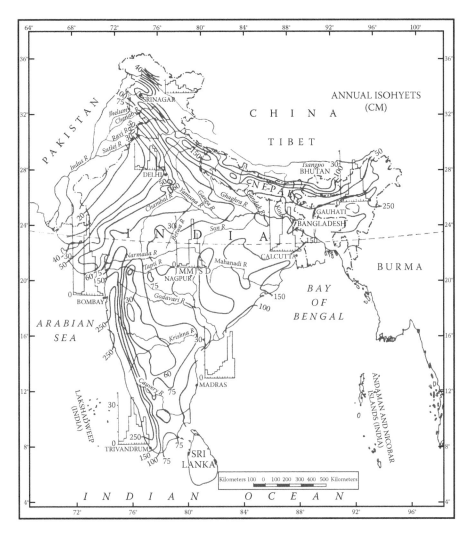

FIGURE 2.6
Annual isohyets. (From Chaturvedi, M.C., *Water—Second India Studies*, Macmillan, New Delhi, 1976.)

coefficient of rainfall variation, representing the percentage plus or minus variation from the mean for the 70% of the year, is shown in Figure 2.7.

The annual precipitation is about 250 cm along the entire West Coast and Western Ghats and over most of the Assam and sub-Himalayan West Bengal. West of the isohyets, joining Porbandar to Delhi and then to Ferozepur, it diminishes rapidly from 50 cm to less than 15 cm to the extreme west. The peninsula has an elongated area with less than 60 cm of rainfall. The normal annual precipitation of 110 cm over the country is slightly more than

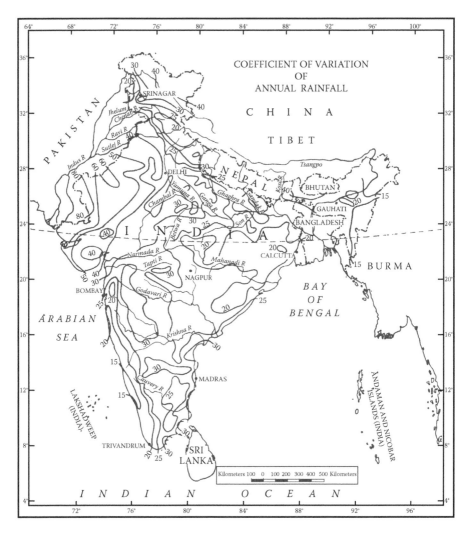

FIGURE 2.7
Coefficient of variation of annual rainfall. (From Chaturvedi, M.C., *Water—Second India Studies*, Macmillan, New Delhi, 1976.)

the global mean of 99.1 cm. However, the point to be noted is the extreme seasonal pattern of rainfall, spatial variation, and yearly variability. The precipitation is confined to essentially the monsoon period of 3–4 months, when essentially about 90% of the rainfall takes place. This is also not a continuous period of rainfall. The precipitation is confined to a few spells every couple of days, and there may be long intervening dry periods. About 50% of the precipitation falls in just 15 days, and over 90% of the river runoff flows in just 4 months. Often, the monsoons may fail.

These characteristics are further demonstrated in terms of annual water surplus and deficits shown in Figures 2.8 and 2.9, respectively. The values are computed by subtracting values of potential evapotranspiration and adding up the annual surplus or deficit separately. The figures show that on the whole, there is a net deficit all over the country. Even in the monsoon months when there is a general surplus, some regions remain drought-stricken. The figures are only indicative, as even in the areas of annual water surplus, there are often seasonal or short-term deficiencies.

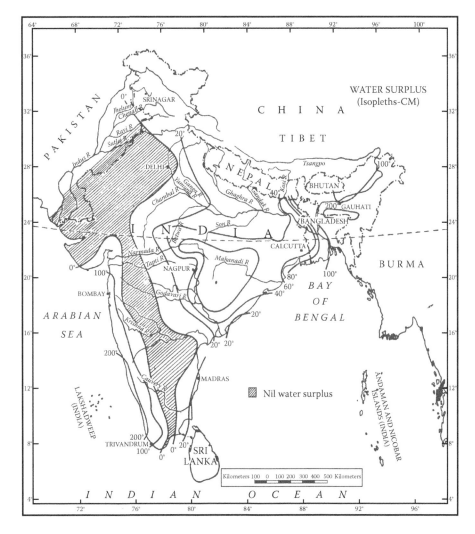

FIGURE 2.8
Water surplus regions. (From Chaturvedi, M.C., *Water—Second India Studies*, Macmillan, New Delhi, 1976.)

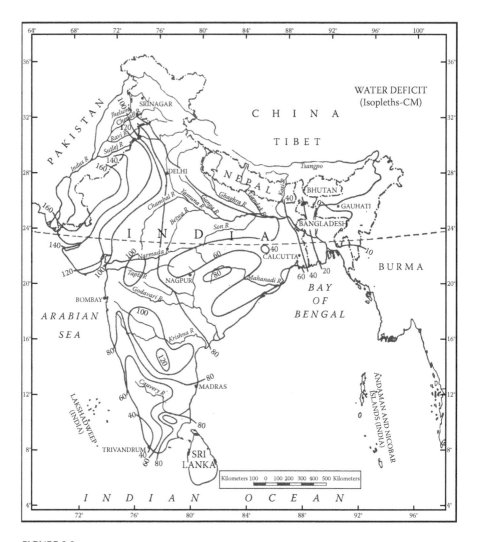

FIGURE 2.9
Water deficit regions. (From Chaturvedi, M.C., *Water—Second India Studies*, Macmillan, New Delhi, 1976.)

The variability of available rainfall is extremely important, as deficiencies during critical periods of crop growth can be disastrous. Besides the considerable geographical and seasonal variation, the rainfall even during the monsoon is most uncertain. There are alterations of heavy to moderate rains and partial or general break when there are no rains. Besides, there is a large annual variation, as shown in Figure 2.6. Furthermore, there are vast variations from month to month. A sketch of the monthly mean and maximum flows of some major rivers at their tail ends is shown in Figure 2.10, which brings out some characteristics of the water environment in India.

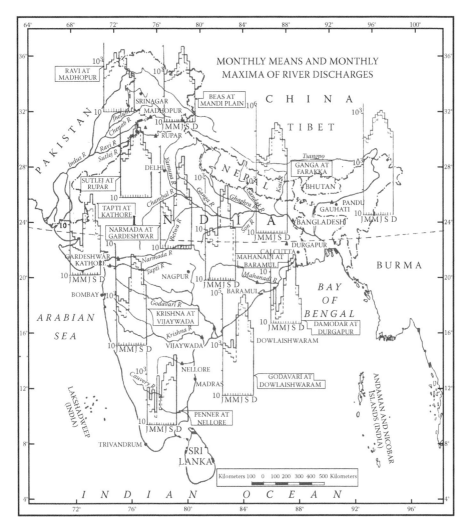

FIGURE 2.10
Monthly mean and monthly maximum discharge. (From Chaturvedi, M.C., *Water—Second India Studies*, Macmillan, New Delhi, 1976.)

2.7.1 Climate Regions

Spatial patterning of the climate phenomena can be very well marked in the country and is reflected in the regional variations. For instance, in the Great Plains, the dry Western Rajasthan is in contrast with the humid eastern section consisting of Assam valley and the Lower Ganga Plain. So secular is the change that within the expanse of over 2000 km, no sharp demarcation can be observed despite tentative attempts at 150, 100, and 75 cm isohyets from east to west.

2.8 Floods and Droughts

The extreme seasonal and spatial variability of water resources of India is dramatically exemplified by occurrence of droughts and floods in different parts of the same river basin at about the same time almost every year. After the scorching heat, monsoons set in the south and northeast. The

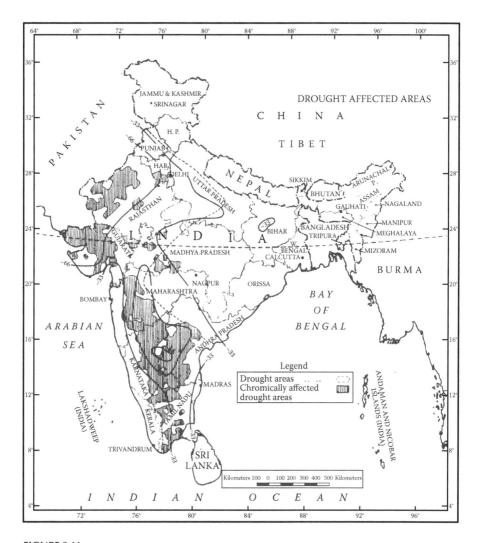

FIGURE 2.11
Drought-affected areas. (From Chaturvedi, M.C., *Water—Second India Studies*, Macmillan, New Delhi, 1976.)

northeastern part of the Ganga–Brahmaputra–Meghna basin starts experiencing heavy floods. The delta region is one of the world's most severely flooded areas. Yet at the same time, the western Ganga basin and several parts of central India continue to suffer severe heat and water scarcity. Very often, the monsoons may be over without these parts receiving adequate rainfall or, in other words, experiencing a drought.

Droughts are extended periods of subnormal precipitation. Whereas the definition of drought is arbitrary, the following definition has been adopted. Semi-arid and arid zones are areas wherein the difference between precipitation received and potential evapotranspiration is less than −33 cm and between −33 and −66 cm, respectively. The drought areas are those that have adverse water balance and 20% probability of rainfall departure of more than 25% from normal. If the probability is less than 40%, then it is a chronically affected drought area. The drought areas are shown in Figure 2.11.

2.9 Groundwater

The 110-cm average annual rainfall provides the country with about 4000 km^3 of water. About one-tenth is used in groundwater recharge. It is distributed very unevenly depending on precipitation, terrain conditions, and lithology, temperature, permeability, etc. Eight groundwater provinces have been identified. The depth of aquifer and quality of groundwater is shown in Figure 2.12.

2.10 River Basins

The river basin is the basic hydrological and environmental unit of land and water. Major river basins are briefly described at the outset to provide a frame of reference for proper physical appreciation of management of water.

The river systems of India have been classified into four groups, namely, (1) Himalayan rivers, (2) Deccan rivers, (3) coastal rivers, and (4) rivers of the inland drainage basin. The Himalayan rivers are formed by melting snow and glaciers and therefore have continuous flow throughout the year. During the monsoon months, the Himalayas receive very heavy rainfall, and rivers swell, causing frequent floods. The Deccan rivers, on the other hand, are rain fed. They therefore also fluctuate considerably in discharge. Many of them are nonperennial. The coastal streams, especially on the west coast, are short in length and have limited catchment areas. Most of them are nonperennial.

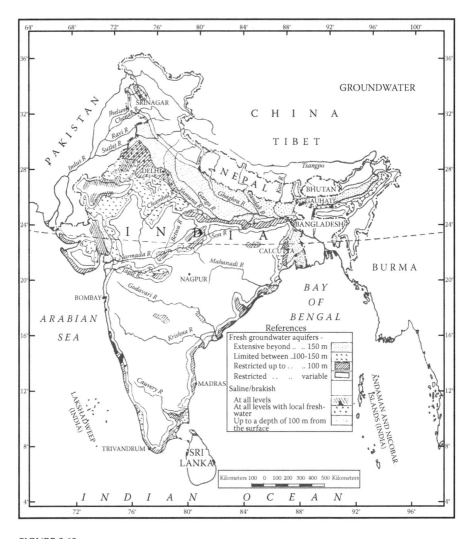

FIGURE 2.12
Depth and quality of groundwater. (From Chaturvedi, M.C., *Water—Second India Studies*, Macmillan, New Delhi, 1976.)

The streams of the inland drainage basin of Western Rajasthan are few and far between. Most of them are of an ephemeral character.

The main Himalayan river systems are those of the Indus and Ganga–Brahmaputra–Meghna system. The important river systems in Deccan are the Narmada and the Tapi, which flow westward into the Arabian Sea and the east-flowing rivers, the Brahmani, the Mahanadi, the Godavari, the Krishna, the Pennar, and the Cauvery, which flow into the Bay of Bengal.

TABLE 2.1

Catchment Area of Basins

Sl. No.	River Basin	Catchment Area (km²)	States Covered in the Basin
1	2	3	4
1	Indus	321,289	J&K, Punjab, Himachal Pradesh, Rajasthan, and Chandigarh
2	Ganga–Brahmaputra–Meghna Basin		Uttar Pradesh, Himachal Pradesh, Haryana, Rajasthan, Madhya Pradesh, Bihar, West Bengal, and Delhi UT
2a	Ganga sub-basin	862,769	Arunachal, Assam, Meghalaya, Nagaland, Sikkim, and West Bengal
2b	Brahmaputra sub-basin	197,316	Assam, Meghalaya, Nagaland, Manipur, Mizoram, and Tripura
2c	Meghna (Barak) sub-basin	41,157	Bihar, West Bengal, and Orissa
3	Subarnarekha	29,196	MP, Bihar, and Orissa
4	Brahmani–Baitarani	51,822	MP, Maharashtra, Bihar, and Orissa
5	Mahanadi	141,589	Maharashtra, AP, MP, Orissa, and Pondicherry
6	Godavari	312,812	Maharashtra, AP, MP, Orissa, and Pondicherry
7	Krishna	258,948	Maharashtra, AP, and Karnataka
8	Pennar	55,213	AP and Karnataka
9	Cauvery	87,900	Tamil Nadu, Karnataka, Kerala, and Pondicherry
10	Tapi	65,145	MP, Maharashtra, and Gujarat
11	Narmada	98,796	MP, Maharashtra, and Gujarat
12	Mahi	34,842	Rajasthan, Gujarat, and MP
13	Sabarmati	21,674	Rajasthan and Gujarat
14	West-flowing rivers of Kachchh, Saurashtra, and Luni	334,390	Rajasthan, Gujarat, and Daman and Diu
15	West-flowing rivers south of Tapi	113,057	Karnataka, Kerala, Goa, Tamil Nadu, Maharashtra, Gujarat, Damn and Diu, and Nagar Haveli
16	East-flowing rivers between Mahanadi and Godavari	49,570	AP and Orissa
17	East-flowing rivers between Godavari and Krishna	12,289	Andhra Pradesh
18	East-flowing rivers between Krishna and Pennar	24,649	Andhra Pradesh

(*continued*)

TABLE 2.1 (Continued)

Catchment Area of Basins

Sl. No.	River Basin	Catchment Area (km²)	States Covered in the Basin
1	2	3	4
19	East-flowing rivers between Pennar and Cauvery	64,751	AP, Karnataka, and Tamil Nadu
20	East-flowing rivers south of Cauvery	35,026	Tamil Nadu and Pondicherry UT
21	Area of North Ladakh not draining into Indus	28,478	Jammu and Kashmir
22	Rivers draining into Bangladesh	10,031	Mizoram and Tripura
23	Rivers draining into Myanmar	26,271	Manipur, Mizoram, and Nagaland
24	Drainage areas of Andaman, Nicobar, and Lakshadweep Islands	8280	Andaman, Nicobar, and Lakshadweep
	Total	3,287,260	

Source: National Commission for Integrated Water Resources Development (NCIWRD), Report, Vol. I, *Integrated Water Resources Development—A Plan for Action*, Ministry of Water Resources, Government of India, New Delhi, 1999.

There are numerous coastal rivers that are comparatively small. Whereas only a handful of such rivers drain into the sea near the deltas of the east coast, there are as many as 600 such rivers on the west coast. Although draining only 3% of the land, they account for as much as 14% of the country's water resources. A few rivers of Rajasthan do not drain into the sea. They drain into salt lakes or get lost in sands with no outlet to the sea.

On the basis of catchments, the river basins of India have been divided into the following three groups: (1) major river basins: river basins with catchment area of 20,000 km² and above; (2) medium river basins: river basins with catchment area between 20,000 and 2000 km²; and (3) minor river basins: river basins with catchment area below 2000 km². There are 12 major river basins, 46 medium river basins, and 6 minor and desert rivers. The Ganga–Brahmaputra–Meghna basin is the largest in the country, receiving waters from an area that comprises about one-third of the total area of the country. It has more than half the population of the country. The second largest basin is that of Indus, even after being divided between India and Pakistan, which covers about 10% of the total area of India. It has become the highest priority in terms of development and represents one of the world's largest integrated developments.

The entire country was suitably subdivided into 20 river basins. These consist of 12 major river basins and 8 other river basins each combining a number of major and minor river basins. This subdivision has been slightly modified, and the country is now considered to have 24 river basins, as shown in Table 2.1. The catchment areas of the basins are shown in Figure 2.2.

The Ganga, Brahmaputra, and Meghna rivers flow into a common terminus before joining the Bay of Bengal; hence, Ganga–Brahmaputra–Meghna is considered as a basin, and the Ganga–Brahmaputra and the Meghna are considered as sub-basins of the Ganga–Brahmaputra–Meghna basin. The Brahmani and Baitarani river systems outfall into the Bay of Bengal, forming a common delta; hence, Brahmani and Baitarani constitute a single basin.

2.11 Water Resources

2.11.1 Natural Flow

Natural (virgin) flow in the river basin is reckoned as water resources of a basin. It is difficult to obtain the natural flow because water resources are already developed to a considerable extent. Therefore, these have to be calculated by making estimates for the impact of these developments on the natural flows. The natural flow at the location of any site is obtained by summing up the observed flow, upstream utilization for irrigation, domestic and industrial uses from both surface and groundwater sources, increase in storage reservoirs (surface and subsurface) and evaporation losses, and deducting return flows from different uses from surface and groundwater sources. Calculations have been made by several agencies. Based on these studies but considering some other data and calculations also, the NCIWRD (1999) has come to a slightly different figure, which is given in Table 2.2, which also shows the utilizable waters, as estimated by the NCIWRD (1999). The author considers that the utilizable waters can be much more, as we will be discussing later in the study. These are, therefore, also given.

Besides the total water flows, an important characteristic is their temporal variation. In view of the precipitation characteristics, there are very heavy flows during the monsoons, tapering off to low flows as the monsoons end and runoff continues to decline. This is the characteristic for all the rivers. The rivers of north, originating from the Himalayas, experience a mild increase as summers set in and the snow starts melting. However, the big change comes only as the monsoons set in. Typical runoff of few rivers of India, representing the northern, central, and peninsular rivers is shown in Figure 2.13.

TABLE 2.2

Water Resources Availability and Utilization[a]

Sl. No.	River Basin	Surface Water			Groundwater			Total Water	
		Mean	Utilizable Commission	Chaturvedi	Replenishable Commission	Artificial Recharge Chaturvedi	Total Chaturvedi	Commission	Chaturvedi
1	Indus	73.31	46.0	46.0	14.29	20.7	34.99	60.29	80.99
2	Ganga–Brahmaputra–Meghna Basin								
2a	Ganga sub-basin	525.02	250.0	250.0	136.47	94.9	231.37	386.47	481.37
2b	Brahmaputra	629.05	24.0	24.0	25.72	15.9	41.62	49.72	65.62
2c	Meghna	48.36	–	–	8.52	–	8.52	8.52	8.62
3	Subarnarekha	12.37	6.81	6.81	1.68	0.6	2.28	8.49	9.09
4	Brahmani–Baitarani	28.48	18.3	18.30	3.35	0.4	3.75	21.65	22.05
5	Mahanadi	66.88	49.99	66.88	13.64	1.5	15.14	63.63	82.02
6	Godavari	110.54	76.30	110.54	33.48	5.8	39.28	109.78	149.82
7	Krishna	69.81	58.00	69.81	19.88	3.3	23.18	77.88	92.99
8	Pennar	6.32	6.86	6.86	4.04	–	4.04	10.90	10.99
9	Cauvery	21.36	19.00	21.36	8.79	1.0	9.79	27.79	31.15
10	Tapi	14.88	14.50	14.88	6.67	–	6.67	21.17	21.42
11	Narmada	45.64	34.50	45.64	9.38	1.4	10.78	43.88	56.92
12	Mahi	11.02	3.10	11.02	3.5	1.4	4.9	6.60	15.92

13	Sabarmati	3.81	1.93	3.81	2.9	0.9	3.8	4.83	7.61
14	West-flowing rivers of Kachchh, Saurashtra, and Luni	15.1	14.98	14.98	9.1	3.3	12.4	24.08	27.38
15	West-flowing rivers south of Tapi	200.94	36.21	36.21	15.55	2.1	17.65	51.76	53.86
16	East-flowing river between Mahanadi and Pennar	22.52	13.11	13.11	12.82	0.6	13.52	25.93	26.63
17	East-flowing rivers between Pennar and south of Cauvery	16.46	16.73	16.73	12.65	4.2	16.85	29.38	33.58
18	Rivers draining into Bangladesh	8.57	NA	NA	—	—	—	—	—
19	Rivers draining into Myanmar	22.43	NA	NA	—	—	—	—	—
	Total	1952.87	690.32	776.94	342.43	158.00	500.43	1032.75	1276.47

Source: National Commission for Integrated Water Resources Development (NCIWRD), Report, Vol. I, *Integrated Water Resources Development—A Plan for Action*, Ministry of Water Resources, Government of India, New Delhi, 1999; Chaturvedi, M.C., *Water Policy*, 3, 297–320, 2001.

[a] In cubic kilometers.

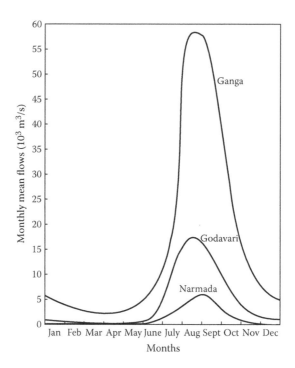

FIGURE 2.13
Typical hydrograph of some major rivers of India.

2.11.2 Groundwater Resource

The existing groundwater regime has been estimated to be 431.9 km^3. This is the sum of the potential from the natural rainfall (342.4 km^3) and the potential due to recharge augmentation from canal irrigation system (89.5 km^3). The dynamic fresh groundwater resource basin-wise is given in Table 2.3. It may be seen that in keeping with the hydrogeological variations in the country, the groundwater development potential also varies widely in different regions.

The limitations of these estimates should, however, be kept in view (NCIWRD 1999). Important limitations are as follows: (1) The accuracy of the assessment of water resources of a river basin made on the basis of river flows measured at a terminal site depends on the accuracy of discharge observations, the reliability of the data on the abstractions upstream, the groundwater withdrawal, the changes in the storages and evaporation losses of the reservoirs, and the return flows/regenerated flow from various uses. Not all these estimates are firm. (2) Major consumption of water in most of the river basins is by irrigation. The state governments do not maintain proper record for other important uses. In many cases, the utilization figures are not available, and varying assumptions have been made to estimate the quantities of utilized water. (3) In most of the cases, the year-wise withdrawal from

TABLE 2.3

Dynamic Fresh Groundwater Resource—Basin-Wise[a]

Sl. No.	River Basin	Total Replenishable Groundwater Resource	Total Replenishable Groundwater Resource Due to Recharge Augmentation from Canal Irrigation	Total Replenishable Groundwater Resource from Normal Natural Recharge
1	2	3	4	5
1	Indus	26.5	12.21	14.29
2	Ganga–Brahmaputra–Meghna Basin			
2a	Ganga sub-basin	171.57	35.1	136.47
2b	Brahmaputra sub-basin	26.55	0.83	25.72
2c	Meghna (Barak) sub-basin	8.52	0	8.52
3	Subarnarekha	1.8	0.12	1.68
4	Brahmani–Baitarani	4.05	0.7	3.35
5	Mahanadi	16.5	2.86	13.64
6	Godavari	40.6	7.12	33.48
7	Krishna	26.4	6.52	19.88
8	Pennar	4.93	0.89	4.04
9	Cauvery	12.3	3.51	8.79
10	Tapi	8.27	1.6	6.67
11	Narmada	10.8	1.42	9.38
12	Mahi	4	0.5	3.5
13	Sabarmati	3.2	0.3	2.9
14	West-flowing rivers of Kachchh, Saurashtra, and Luni	11.2	2.1	9.1
15	West-flowing rivers south of Tapi	17.7	2.15	15.55
16	East-flowing rivers between Mahanadi and Godavari	[18.8]	[5.98]	[12.82]
17	East-flowing rivers between Godavari and Krishna			
18	East-flowing rivers between Krishna and Pennar			
19	East-flowing rivers between Pennar and Cauvery	[18.2]	[5.55]	[12.65]
20	East-flowing rivers south of Cauvery			

(continued)

TABLE 2.3 (Continued)

Dynamic Fresh Groundwater Resource—Basin-Wise[a]

Sl. No.	River Basin	Total Replenishable Groundwater Resource	Total Replenishable Groundwater Resource Due to Recharge Augmentation from Canal Irrigation	Total Replenishable Groundwater Resource from Normal Natural Recharge
1	2	3	4	5
21	Area of North Ladakh not draining into Indus		Not Assessed	
22	Rivers draining into Bangladesh		Not Assessed	
23	Rivers draining into Myanmar		Not Assessed	
24	Drainage areas of Andaman, Nicobar, and Lakshadweep Islands		Not Assessed	
	Total	431.89	89.46	342.43

Source: National Commission for Integrated Water Resources Development (NCIWRD), Report, Vol. I, *Integrated Water Resources Development—A Plan for Action*, Ministry of Water Resources, Government of India, New Delhi, 1999.

[a] In cubic kilometers per year.

groundwater was estimated on the assumption of linear variation between the state-wise draft by the Irrigation Commission (GOI 1972) for the years 1967–1968 and by the Central Ground Water Board (CGWB) for the years 1983–1984, by interpolating for other years. (4) The return flow for irrigation from surface water resources was assumed at 10%–20% of use, and the return flow from groundwater was not accounted for. The return flows for domestic and industrial uses were assumed at 80% of the use. The irrigation efficiency of surface water was estimated to be in the order of 30%–40%, and efficiency of groundwater use was about 70%–75%. (5) The evaporation losses were assumed at 20% of the annual utilization wherever project authorities did not maintain the records of evaporation losses.

2.11.3 Static Groundwater Resource

The static fresh groundwater resource is considered as groundwater available in the aquifer zones below the zone of water level fluctuation. Preliminary studies indicate that in alluvium, groundwater can be exploited down to 450 m, as in the Indo-Ganga valley. The coastal aquifers are also having similar depth range of groundwater availability. Inland river basins in the country have recorded shallower depth within the range of 100–150 m. In Gondwana Territories of Maharashtra and Andhra Pradesh, groundwater can

TABLE 2.4

Static Fresh Groundwater Resource—Basin-Wise[a]

Sl. No.	River Basin	Static Fresh Groundwater Resource		
		Alluvium/ Unconsolidated Rocks	Hard Rocks	Total
1	2	3	4	5
1	Indus	1334.9	3.3	1338.2
2	Ganga–Brahmaputra–Meghna Basin			
2a	Ganga sub-basin	7769.1	65	7834.1
2b	Brahmaputra sub-basin	917.2	0	917.2
2c	Meghna (Barak) sub-basin	101.3	0	101.3
3	Subarnarekha	10.1	0.7	10.8
4	Brahmani–Baitarani	40.1	3.3	43.4
5	Mahanadi	108.4	11.3	119.7
6	Godavari	36	23.4	59.4
7	Krishna	13.6	22.4	36
8	Pennar	3.9	7.2	11.1
9	Cauvery	39.1	3.3	42.4
10	Tapi	4.3	3.2	7.5
11	Narmada	13.8	4.6	18.4
12	Mahi	9.7	2.9	12.6
13	Sabarmati	25.5	2.7	28.2
14	West-flowing rivers of Kachchh, Saurashtra, and Luni	103.1	10.1	113.2
15	West-flowing rivers south of Tapi	5.4		11.2
16	East-flowing rivers between Mahanadi and Godavari	[34.4]	[6.9]	[41.3]
17	East-flowing rivers between Godavari and Krishna			
18	East-flowing rivers between Krishna and Pennar			
19	East-flowing rivers between Pennar and Cauvery	[63.1]	[2.9]	[66]
20	East-flowing rivers south of Cauvery			
21	Area of North Ladakh not draining into Indus	Not Assessed		
22	Rivers draining into Bangladesh	Not Assessed		
23	Rivers draining into Myanmar	Not Assessed		
24	Drainage areas of Andaman, Nicobar, and Lakshadweep Islands	Not Assessed		
	Total	10,633.00	179.0	10,812.0

Source: National Commission for Integrated Water Resources Development (NCIWRD), Report, Vol. I, *Integrated Water Resources Development—A Plan for Action*, Ministry of Water Resources, Government of India, New Delhi, 1999.

[a] In cubic kilometers.

be extracted up to depth range of 175–250 m. In hard rock terrain, the availability of groundwater increases steadily up to around 100 m where after the frequency of water-yielding fractures diminishes except in sporadic cases.

The general range of yield in the unconsolidated formation is from 200 to 350 m³/h. Some of the prominent water-bearing strata in these formations are piedmont alluvial plain, glaciolacustrine deposits, inland river flood plains, and coastal alluvium. In the Indo-Ganga alluvial, tubewells were constructed up to the depth of 600 m with a yield prospect of 400 m³/h. In the semi-consolidated formations, the groundwater productivity is fairly good. The lathi formation in western Rajasthan has recorded the highest yield of 450 m³/h at a depth of 544 m. The general yield range is 60–200 m³/h for depths of 250–400 m. The groundwater prospects in the fractured rock formations in peninsular India are nonhomogeneous and site specific. The yield of borewells tends to decrease with increasing depth because of reduction in weathering, closure of joints and fracture openings, and lack of interconnection between fractures. However, in tectonically weaker zones in hard rock formations, the wells have recorded a good yield. The average range of well yield in these formations is from 50 to 150 m³/h at favorable locations.

An assessment of the quantum of static groundwater resource available in the country has been carried out by the CGWB on the basis of the depth of availability of groundwater and the productivity of deeper aquifers. The estimate of static yield has been made district-wise in different states of the country. The total estimated groundwater resource is 10,812 km³. The details basin-wise are given in Table 2.4.

2.11.4 International Perspective

It may be in order to have a comparative international perspective of water availability. Some figures are given in Table 2.5. It must, however, be stated at

TABLE 2.5

International Perspective

Country	Annual Internal Renewable Resources (km³)	Water per Land Area (m³/km²)	Water per Capita (m³)
Canada	2901	290,944	98,462
Brazil	6950	816,494	42,957
Russian Federation	4498	263,426	30,599
United States	2478	264,659	9413
China	2812	292,917	2292
India	2085	700,840	2282
World	41,022	301,988	7176

Source: World Resources Institute, *World Resources 1996–97*, Oxford University Press, New York, 1996. World Bank, *World Resources 1996–97*, Oxford University Press, New York, 1997.

the outset that physiographic, climatic, demographic, and economic factors make these figures rather difficult to comprehend.

The annual water availability in India is estimated at about 4000 km³, which includes 2000 km³ of water that is brought into the country by rivers outside India. The annual internal renewable resources are estimated at 2085 km³.

2.12 Climate Change

A serious issue of climate change has arisen. United Nations Framework Convention on Climate Change has been established to stabilize greenhouse gas concentrations. Some salient aspects of the information provided in the Indian National Communication to it are as follows.

Significant increase in the order of 0.4°C in the past 100 years in the annual global average surface air temperatures has already been observed. Whereas the annual average monsoon rainfall at the all-India level for the same period has been without any trend and variations have been random in nature, increase in monsoon seasonal rainfall has been recorded along the west coast, north Andhra, and northwest India (+10 to +12% of normal/100 years), and decreasing trends have been observed over east Madhya Pradesh and adjoining areas, northeast India, and parts of Gujarat and Karalla (−6% to 8% of normal/100 years). Using the latest modeling practices, marked increase in seasonal surface temperature is projected into the twenty-first century, becoming conspicuous after the 2040s. Climate projections indicate increases in both maximum as well as minimum projections over the regions south of 25°N; the maximum projection is seen to increase by 2–4°C during the 2050s. In the northern region, the increase in maximum temperature may exceed 4°C all over the country, which may increase further in the southern peninsula. Little change in monsoon rainfall is predicted up to the 2050s at the all-India scale level. However, there is an overall decrease in the number of rainfall days over a major part of the country. This decrease is greater in the western and central parts (by more than 15 days), whereas near the Himalayan foothills (Uttranchal) and northeast India, the number of rainy days may increase by 5–10 days. Increase in rainfall intensity by 1–4 mm/day is expected all over India, except for some small areas in northwest India, where intensities may decrease by 1 mm/day.

Regarding water resources, using the Soil and Water Assessment Tool water balance model for the hydrological modeling of different river basins in the country, in combination with the outputs of the Had RM2 regional climate model, preliminary assessments have revealed that under the IS 92a scenario, the severity of droughts and the intensity of floods in various parts of India are likely to increase. Furthermore, there is a general reduction in the quantity of available runoff under the IS 92a scenario. River basins of Sabarmati and Gujarat are likely to experience acute water-scarce conditions. River basins of Mahi, Pennar, Sabarmati, and Tapi are likely to experience constant water scarcity and shortage. River basins of Ganga, Narmada, and Krishna are likely to experience seasonal or regular water-stressed conditions. River basins of Godavari, Brahmani, and Mahanadi are projected to experience water shortage only in few locations.

Note

1. The chapter follows from Chaturvedi (2011a).

3

Development and Management of India's Waters—Overview

3.1 Historical Perspective

Attempts to live according to the availability of water and develop and manage water have been made from the earliest times. Settlements took place along rivers, lakes, and ponds. Dug wells and tanks were constructed for drinking water and irrigation. Fields along rivers were inundated by transferring water to irrigate the crops. Canals were constructed for irrigation. Some large storage works were undertaken with the support of benevolent kings, but they were rare. Management of water has been important ever since Ashoka's edicts in the fourth century BC were brought out.

With the influx of the Muslim rulers in the arid northwestern region, new experiences and technologies from Central Asia were introduced. Ghiyasuddin Tughlaq (1220–1225) is credited to be the first ruler who encouraged these activities. However, it is Firuz Tughlaq (1351–1386) who is considered to be the greatest canal builder before the developments undertaken by the British in the nineteenth century. Hissar was irrigated by the Rajab wa and the Ulugh-Ichan from the river Yamuna, and the Firuz-shahi from Sutlej. One canal was from Ghaggar and one from Kali in the Doab of Yamuna near Delhi. Furthermore, there were a number of smaller canals. In Multan, the river was dredged by the state, but canals were dug and maintained by the local population on fear of death and exile. Firuz Tughlaq took water tax (*haqq-i-shurb*) in Haryana, which was one-tenth of what it produced.

These practices were continued by the succeeding Mughal rulers. Babar refined the *araghatta* to the modern Persian wheel. Shah Jahan's reign was marked by the construction of several canals for irrigation. Nahr-Ul Faiz was carved out of an old canal to which another 78 miles was added, making it over 150 miles long, taking off from Yamuna River at the point where it leaves the hills to join the parent river in Delhi. Another canal, about 100 miles long, took off from Ravi River at Rajpur. Beginning at the same point, one canal ran to Pathankot, another to Batala, and a third to Patti Haibatpur. Traces of other canals are found all over the Indus plains down to the delta.

Financial support was also provided for development of irrigation. For example, in the early Sultanate period, in the reign of Alauddin Khilji (1296–1316), state credit in the form of *taqavi* (credit) was offered to the peasants. Again, the famine in the first decade of the reign of Muhammad Tughlaq (1325–1351) forced him to take steps to provide *taqavi* to the peasants to dig wells, both to safeguard existing crops and extend cultivation. This was also a way for the state to eliminate usury and make inroads into peasant's granary, demanding in return a part of the produce. This policy was soon institutionalized and became a routine administrative practice until the end of the Mughal Empire, and was even in force under the British administration.

During the Mughal period, the consolidation of petty peasant production took place, perhaps because of the limitations of the scope of large irrigation works and because of the continuation of small irrigation works suitable for small holdings. The production process was primarily based on family labor. However, there were a few large farms, though the latifundia or plantation have not been found. The nobility were, in essence, rent receivers. The caste system prevalent in India made it possible for the entire community of cultivators to meet their labor requirement from the menial castes, who did not have the right to own and cultivate land. In this way, Indian feudalism solved its labor problem, but further intensification of agriculture was not possible. Thus, the only way cultivation could be expanded and agricultural output increased was through capital investment in the form of irrigation devices and high-value crops. Much development in this context did not take place as the State took little interest. Agricultural technologies continued to be in the hands of the peasants, who was hampered in their efforts to bring about more efficient changes.

Moreover, in the south, people from early times constructed tanks. State intervention began at least as early as 300 AD, as brought out by the construction of the Grand Anicut by the Cholas to provide irrigation from Cauvery River in the delta area. The Viranam tank of south Arcot, a freshwater lake in the State that is over 10 miles long and 3.5 miles in width, was constructed by Rajendra Chola (1011–1037 AD). Support to people in construction of tanks was also provided by the rulers. Irrigation is considered to be the major reason for the growth and expansion of the Vijayanagara Empire as referred to by several Portuguese travelers. The state actively patronized the construction of tanks and reservoirs, for example, the Maday Lake, said to be about 10–15 miles long. Other regions also saw involvement of princes in development of irrigation. Tanks were built all over the peninsular region. Other indigenous devices were also used. However, it should be noted that, until the beginning of the nineteenth century, only 3%–7% of the cultivated area was irrigated in most parts of south India, except in Tanjore, where about 50% of the land was irrigated through the Grand Anicut.

The early developments have been considered in some detail for historical interest. The population, compared with current and future, was very small, estimated to be about 100 million until the end of the early seventeenth century. It was still 253 million, of the entire subcontinent, at the time of the first census carried out by the British administration in 1881 (Thapar 1977). The economy was based on sustenance agriculture. The technology of development was elementary, which did little to provide adequate and assured water supplies. Even at the height of the Moghul Empire in the seventeenth century, the actual extent of public irrigation works was fairly limited. Madison (1971) states that, "these were unimportant and probably did not cover more than 5% of the cultivated land of India." Irrigation did little to minimize the ravages of famines, which were frequent and disastrous.

3.2 British Period

3.2.1 Development

As the Mughal Empire was crumbling and the British stepped in, there was a long period of turmoil. Although British role in the sociopolitical sphere started from the mid-eighteenth century, the British hegemony was established up to Sutlej in the Indo-Gangetic basin only by 1818. It was a strange development, a trading company gradually emerging as a successor of the vast decaying Mughal Empire. The British India, yet, consisted mainly of Bengal, Bihar, a tract to the north of the Ganges running beyond Delhi up to Sutlej, the coastal Carnatic in the south, and a region around Bombay. Administration of the British India at that time was in terms of the three Presidencies of Madras, Bengal, and Bombay.

The India that the British inherited was very different from the India of the Great Moguls, in a large measure due to the year 1818. The British found a country in ruins. Not only did they encounter dismantled fortresses and deserted palaces, but also canals run dry, tanks or reservoirs broken, roads neglected, towns in decay, and the whole region depopulated. The lengthy war and famines in between had ravaged the country and completely broken down the social system (Spear 1978).

The British engineers had no experience of irrigation works. One of their early activities was renovation of the Western Yamuna Canal, constructed in the fourteenth century, attributed to Firuz Shah Tughlaq, which had fallen under disuse. The activities started in dramatic fashion. A severe famine gripped the country in 1820. The region around Delhi, the Yamuna basin, was badly affected. Some relief was available to people from an existing Yamuna canal, which had essentially been constructed to provide water to the hunting areas of the Mughal kings. However, it also provided some benefits to the people, but it had become dysfunctional on account of neglect during the gradual erosion of Mughal power. An attempt was made by the now ruling British powers, a trading company, to restore the canal. It earned considerable appreciation of the people for the emerging royal role of the East India Company, which had come to be called Company Bahadur (valiant in local language). It was reopened in 1820. Rehabilitation of the Eastern Yamuna Canal (EYC), built in earlier times, was next undertaken, which was opened in 1830. Besides the political advantages, equally significant for the trading company was the excellent return that the investment provided: about 14% (Buckley 1893). The engineer who undertook the construction of the EYC was Lt. Proby T. Cautley, who was to introduce the modern canal irrigation in India later.

Most of the early British schemes were, in fact, rehabilitated and extended versions of indigenous works found in various parts of the country. Also functioning, when Tanjore was ceded to the British in 1801, was the work that perhaps was the most impressive, and which, when rehabilitated, was to form part of Sir Arthur Cotton's extensive irrigation works in Madras—the 300-m "Grand Anicut." This work, which spanned Cauvery, has been ascribed to the Chola kings. At one time, it irrigated an estimated 0.25 million hectares (Mha).

Cotton also constructed the Godavari Anicut and the canal system. This diversion system started in 1846, irrigating 0.4 Mha, and contributed much to the mitigation of severe famines in the area. Similarly, Cotton suggested an anicut across the Krishna and a canal system. Its construction was started in 1852 and was completed in 1853. It irrigates about 0.5 Mha. All these works established Lt. General Cotton as a great hydraulic engineer and the British Government knighted him.

Cautley undertook even more impressive work in the north. With the attraction of the financial benefits of irrigation and the experience and confidence gained with the restoration of the EYC, Cautley drew up plans to construct the Ganga Canal [later called Upper Ganges Canal (UGC), as another canal, lower down, was constructed]. The idea was to divert the Ganges waters as the river debauches into the plains at Hardwar to irrigate the land between Ganga and Yamuna, where several famines had occurred. An inundation canal already existing at the site was a great inspiration. The project was conceived in 1840, but the Sikh wars interrupted the work. However, this also gave Cautley an opportunity to see the numerous inundation canals in the adjoining Indus basin, which developed considerable confidence. He went to Italy to see the irrigation works developed on River Po. The UGC was completed in 1854 and was a great success, technologically and financially. It was much bigger than any existing canal, irrigation or navigation, in the world at that time. It was designed to carry 191 m^3/s (6750 ft^3/s). It irrigated 488,000 ha using 3700 km of canals and distributaries. Navigation was an important function of canals being developed in Europe at that time, and navigation facilities were also provided in the UGC. The canal cost 2.15 million pounds (about half the cost of constructing Taj Mahal) (Schwarz et al. 1990).

With these works, a new era of canal irrigation started in India and even in other colonies of the British Empire, such as in Egypt. Certain political developments also supported these developments. Ranjit Singh, the ruler of Punjab, died in 1839. Two rulers and a whole posse of leading chiefs of that area died violently in the next six years. After a bloody but brief Sikh war, the region was brought under British control as a "sponsored state" under the direction of Sir Henry Lawrence. However, within weeks of his departure in 1848, a revolt began and a second bloody but brief Sikh war of 1848–1849 followed. This time, the area was annexed by the British as the British rulers considered that the area was too near the frontier for any further risks to be undertaken. A political–administrative culture, which was different than in the vast Gangetic plains or the peninsular coastal regions already under British control for about a century, was adopted. It was a sort of direct rule under the control of the Governor, Henry and John Lawrence, and the direction of the Viceroy Dalhousie himself. Thus, the following major water resources developmental activity during the British period was focused in the Indus basin. It started with renovation of old works and gradual undertaking works of increasing magnitude, which would lead to one of the world's major irrigation development shortly.

Attracted by the profitability of the first major irrigation works and basing trust on the views of Cotton, who advocated linking the rivers from considerations of irrigation and navigation, which had been found to be very profitable in England, two private companies planned the development of the water resources on a grandiose scale. Their aim was to link Karachi via Kanpur, Calcutta, and Cuttack to Bhatkal, Mangalore, and Madras. However, all they were able to achieve was a series of disconnected waterways such as the Midnapore Canal. The venture ended in a failure and had to be taken over by the government. This episode led to a policy review, and it was decided that, in the future, the government would undertake irrigation development. These two companies nevertheless left their mark on irrigation financing. For their capital, they had depended on public loans. This method of raising capital came to stay. The government adopted this practice in 1867, in respect of works, which promised a minimum net return.

A new chapter in water resources engineering and irrigation in India was introduced when the government decided in 1866 to take upon itself the planning and construction of completely new canal systems with weir-controlled supplies. This started the development of the new phase of irrigation in India, which continues with increasing improvements. The first project to be conceived independently of previous inundation canals was that of Sirhind canal, and it contributed new experiment in planning and design. Thereafter, a number of projects were undertaken. These included major canal works such as the Lower Ganga, the Agra and Mathura Canals, and the Periyar irrigation works, and smaller works such as the Pennar River Canals (Andhra Pradesh), the Hatmati Canal (Gujarat), the Ekruk Tank, and the Lakh Canal (Maharashtra). Some other projects were also completed on the Indus system. These included the Lower Swat, the Lower Sohag and Para, the Lower Chenab, and the Sidhani Canals.

The recurrence of drought and famines during the second half of the nineteenth century necessitated development of irrigation to give protection against the failure of crops and to reduce large-scale expenditure on famine relief. As irrigation works in low rainfall tracts were not expected to meet the productivity test, they had to be financed from current revenue. Some protective works were, therefore, also undertaken.

The total irrigation from all sources in 1900 was 13.4 Mha of which public works accounted for 56%. The gross area sown was 82.2 Mha of which about 16% was irrigated. Source wise, canals irrigated 45% of the area, wells 35%, tanks 15%, and other sources 5%. Punjab was leading India in irrigation. Of the 11.2 million acres irrigated by major canals in the Indian Empire (including princely states) in 1900–1901, 4.6 million acres was in Punjab (Madras ranked second with 2.9 million acres and United Provinces third with 1.9 million acres).

The major focus remained the Indus basin, and several works in the Indus basin were undertaken. One major irrigation work in the Ganga basin was development of irrigation from R. Sarda, in 1928, which is a tributary of Ganga. The headworks are prominently in the British India, although a portion of land had to be obtained from Nepal for the construction of the eastern end of the headwork. It was duly paid, but Nepal engineers harbor resentment about ceding the right to use the shared waters to India (Gyawali 2001).

The idea of building Bhakra Dam, a storage project on R. Sutlej, had been developed early in the twentieth century, but disputes between Punjab and Sind delayed implementation. Thus, canal irrigation remained the central activity of water resources development during the British rule.

The position of irrigation at the time of partition in India and Pakistan is given in Table 3.1.

3.2.2 Development Policy

A constellation of technological, policy, financial, and practical factors dictated the policy of development of the water resources in the British period. It is important to understand them as even though all these factors have changed, basic analysis of the policy of development and management of water has not been undertaken and the historical colonial British policy momentum continues.

There was hardly any industrial development or urbanization in the British period. Development of water was only in the context of irrigation, and water resources development became synonymous with irrigation.

The irrigation development under British rule began with the renovation, improvement, and extension of some of the major existing works. Thus, the early technology, with some improvements, continued to be followed. Temporary obstructions, as in the earlier inundation canals, were used to divert the low flows after the monsoons for providing irrigation to crops during the ensuing period (called *Rabi* in the local language). This was an annual expenditure, and construction of the headworks was not easy or economical. A large canal command was, therefore, tried to be covered from one headwork. Only part of the canal command, about 30%, was provided irrigation from several considerations. First, the basic objective was to provide irrigation

TABLE 3.1

Net Sown Area and Irrigated Area in India and Pakistan on the Eve of Partition (Average for 1944–1945 to 1946–1947)[a]

| | Net Sown Area | Net Irrigated Area | Column 3 as Percentage of Column 2 | Area Irrigated By | | | | | |
				Government Canals	Private Canals	Canals (Total)	Wells	Tanks	Others
1	2	3	4	5	6	7	8	9	10
India	98.5	19.4 (100)	19.7	6.3 (32.5)	1.9 (9.8)	8.2 (42.3)	5.3 (27.3)	3.3 (17.0)	2.6 (13.4)
Pakistan[b]	18.3	8.8 (100)	48.1	6.8 (77.2)	0.2 (2.3)	7.0 (79.5)	1.3 (14.8)	– (–)	0.5 (5.7)
Undivided India	116.8	28.2 (100)	24.1	13.1 (46.5)	2.1 (7.4)	15.2 (53.9)	6.6 (23.4)	3.3 (11.7)	3.1 (11.0)

Source: Government of India, *Report of the Second Irrigation Commission*, Ministry of Irrigation, New Delhi, 1972.

Note: Figures in parentheses give percentages to the total.

[a] In million hectares.

[b] Figures for Pakistan are estimated figures.

distributed over a large area to minimize the threat of famines. Improvement of productivity was not the central objective. Second, to keep costs low, drainage was not provided, and, therefore, coverage of a large area was also considered appropriate to minimize the possibility of waterlogging.

Three waterings were proposed to be provided, as per existing practice of irrigation at that time. Because the river supplies continuously dwindle after the monsoons, and there was no provision for storage, there was not enough water to provide adequate second and third waterings or even provide water at all to all the areas for the third watering. The function of the canal was to provide $0.028 \text{ m}^3/\text{s}$ (1 ft^3/s) of water at the outlet. Further distribution among the farmers was left to the villagers, as administratively it was not possible to manage at the village level. There was also not much concern about agricultural developments, which will require sophistication about supply of water. The poor farmer could not be expected to pay much, and, therefore, provision of water, as per earlier indigenous inundation canals, was considered adequate. Thus, the objective of canal irrigation was to provide irrigation to stabilize the subsistence agriculture. It was not meant to be an agent of agricultural development.

The environmental implications were not well understood at that time. There were concerns that canal irrigation may introduce waterlogging and malaria. The canal design was suitably modified to minimize these hazards. The adverse environmental impacts on account of reduced low flows in the river were not taken into account because domestic and industrial water supplies were not developed. Almost all the low season flows were diverted, and the river, after the canal headwork, almost became dry. The return flows from irrigation gradually recharged the river. As sufficient discharge became available, it was again diverted to irrigate more land, and the river again became dry. A formidable consequence of the policy has been that, with the exploding urbanization after Independence and increasing development of industry, which is still minimal and essentially in the unorganized sector, the rivers have been converted into open drains.

There was neither development of storage projects nor much development of groundwater. Therefore, there was no possibility of providing adequate and reliable water supplies for irrigation even if it was considered desirable. The policy of stabilizing the sustenance agriculture preempted such an approach. Indeed, even in the development of groundwater in the public sector, where adequate, timely, and reliable water supplies could be provided, the policy of providing an apology of irrigation continued. There was little concern for such welfare activities as flood mitigation. Hydropower was left almost undeveloped.

The physiographic–hydraulic conditions of the Himalayan Rivers are basically different from those of the peninsular rivers, and this has introduced some characteristic differences in development of water in the two regions. The Himalayan Rivers are perennial. They debouch in the plains from steep mountains and then flow in the flat alluvial plains, meeting the ocean finally where a big delta is formed. The peninsular rivers are not perennial. Their descent from the Western Ghats is not so dramatic, they do not traverse alluvial plains, and their deltas are not so big. Storages were difficult to build in the young and fragile Himalayan Mountains. Thus, the general principle in the Himalayan Rivers was that diversion canals were constructed on the rivers just as they debouched in the plains. As recharge of the rivers took place and enough water became available, they were again diverted for irrigation. Generally, no consideration was kept for maintaining certain minimum flows from environmental considerations or demands other than those of irrigation, such as domestic and industrial demands, because there was little urbanization or industrialization. Besides irrigation, other uses of water, namely, navigation, flood control, drainage, hydroelectric development, and even water supply for drinking purposes, did not get any importance.

Moreover, as we emphasized at the outset, political factor made a great difference in development of water resources in the north and south during the British period. Highest emphasis was laid on development of irrigation in the Indus basin. It could also be undertaken without any political constraints as the entire region was under British control. Even if there were princely states, they were totally subservient. In the peninsular region, the conflicts of the princely state of Mysore and British Indian Madras State over development of river Cauvery arose as far back as 1907.

Thus, water resources development meant irrigation. Irrigation meant providing an apology of water to stabilize the sustenance agriculture. Even for irrigation, the colonial government was content with supplying water up to the canal outlet for Rabi (winter period in local parlance), leaving the rest of the task of conveyance and distribution to the farmers themselves. Activities such as land-leveling, drainage, lining of water courses, and so on, which demanded public participation, were also not considered possible by a foreign government. They were not within the capability of the poor peasantry and were, therefore, neglected. Thus, in short, a policy of inefficient extensive irrigation through low flow diversions to stabilize the sustenance agriculture evolved from a convergence of social, political, economic, technological, and financial considerations. It is significant to note that the per capita food availability continued to drop sharply until Independence, as shown in Figure 3.1.

3.2.3 Institutional Characteristics and Governance

In the British period, India was Her Majesty's Government. It was managed from the perspective of maintaining the political control and maintaining law and order, just as the earlier Mughal emperors had ruled, except that the emperor or empress, who was white, set in a throne far away in an island and ruled India through a handful of British officials. In the early phases of the British rule, most of the activities were undertaken by the army. Gradually, a civil service was instituted. The senior positions were held by British officers. They engaged a vast number of local people in subordinate positions to do their command. India was a unique colonial government. Surprisingly, people accepted the humiliating subjugation for a long time.

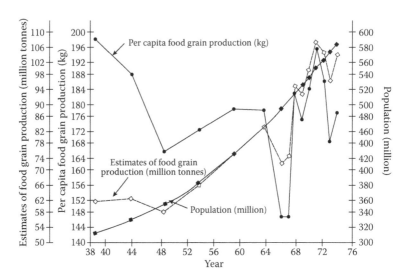

FIGURE 3.1
Population and food dilemma of India in early years.

Water resources development also started in this perspective. The development and management were undertaken by the military officers, and gradually a civil service was established. The officers came from England, but as the canals were constructed, Roorkee College was established in 1847 at Roorkee, near Hardwar, the headworks of the Ganga Canal, as its construction was undertaken. It was renamed Thomason College of Civil Engineering in honor of Sir James Thomason, who was the first Chief Engineer and later Lt. Governor of the United Provinces. It was founded to train the supporting junior engineers for the public works department activities, and the technical capability was poor. This continued until 1946, just before Independence, the author being a student of the last batch of 1946. It was upgraded and titled University of Roorkee in 1948 after Independence (upgraded as Indian Institute of Technology, Roorkee, in 2001).

3.2.4 Scene Just before Independence

As discussed above, water resources development meant irrigation. Drinking water supplies or sanitation was totally ignored. Canal irrigation was the technology, and stabilization of the subsistence agriculture was the policy in British India. Some changes were, however, taking place in the period toward the end of the British regime.

Highest priority for irrigation in British India was in the Indus basin, as noted, on account of two factors complementing each other—political and technological. As canal irrigation was developed in the Indus basin, it was considered that storages should be constructed. The concept of Bhakra Dam on R. Sutlej emerged, and preliminary proposals were prepared. However, conflict between the upstream province Punjab and the downstream province Sindh developed. Attempts were made to resolve the conflict by appointing a Committee under the Chairmanship of Sir Bose. However, before further action could be taken, India achieved Independence.

An important development, though on a small scale, was also taking place toward the end of the British period. Some hydroelectric projects were undertaken on the UGC and Sarda Canal. The canals had a small slope, which was in conformity with the land slope and river slope in the plains, but was much smaller than the slope of the ground in the upper reaches of the Himalayan Rivers. Therefore, the canal design in these reaches led to introduction of falls after certain lengths of the canal. The principle of design was to construct a small hydroelectric power house at the location of a fall or a constellation of falls grouped together to utilize the potential energy of the falls on the canals. The technology, environment, and management culture shaped the policy of managing the small amount of hydroelectric energy and water that became available. There was hardly any industrialization or industrial capacity. Tube wells were developed, first and foremost, to utilize this hydro energy. Even the construction of the necessary pumps and electric motors had to be undertaken in workshops established by the government, specifically for this purpose. The development of the tube wells in the fields and management of water from tube wells became the responsibility of the government. The management of water, therefore, was undertaken along the lines of the management of canal waters, which had carried on for a long time and which still dominated the scene for this purpose. The reverse connection may be noted. The projects were undertaken not because tube wells were required for irrigation; it was the other way around. Water supplies from tube wells could have provided assured timely irrigation, but instead, the policy adopted was in conformity with the existing canal irrigation policy—extending the apology of irrigation.

Construction of dams had become attractive after extensive developments in the United States and the Bhakra Dam on R. Sutlej. In the Gangetic Plains, storage projects were also investigated in the Himalayas. Nayar Dam, a 200-m-high dam, was investigated. However, the foundations were found to be unsafe, and the project was given up. Emphasis was laid on completion of the ongoing canal power houses in the Himalayas.

3.3 Development after Independence

3.3.1 Development—Early Stages

A new era, expressed by Nehru's tryst with destiny speech in the Parliament on the occasion of achievement of independence, started after independence in 1947. Planned development was undertaken with the First Five-Year Plan

(1951–1956), and water resources development was given the highest importance. Water resources development was, however, synonymous with irrigation and multipurpose projects. As the major projects were being planned, work was started on many projects, which had been envisaged even before Independence. Several run-of-canal hydroelectric schemes under which the falls in the upper reaches of the major canals were proposed to be converted into small hydropower stations had found favor with the completion of some earlier ones, and several others had been undertaken in British India, as mentioned earlier. They were continued and more were planned. These included Ganguwal on Rupar Canal, Pathri on UGC, and Khatima Power House on Sarda Canal. Construction of these had been started before attainment of Independence.

Water resources development had been transformed in terms of multipurpose projects, which had been undertaken in large numbers in the United States from the time of the Depression, which transformed their economy. Some far-sighted engineers (led by Dr. A.N. Khosla) had been pursuing a major dam on Sutlej, but it was kept in cold storage. With the Independence, activities on undertaking storage projects started in real earnest. Some of the prominent and early ones were Bhakra–Nangal, the Damodar Valley, the Hirakud Dam, and Rihand Dam. Bhakra was a 226-m (700-ft) high concrete gravity dam undertaken by the Punjab State. Damodar was undertaken on lines of Tennessee Valley Authority and represented integrated land and water development, launching initiation of soil conservation service. Hirakud was undertaken by the Government of India on Mahanadi. In the Gangetic Plains, focus was on development of dams in the Vindhyan region, in Uttar Pradesh, after the experience of the foundation problems in the Himalayas. Kosi Dam was, however, investigated on river Kosi in view of the enormous human misery caused by the floods, but had to be given up as a result of the inadequate cooperation of the Nepal government. Instead, Kosi Canal was constructed. Investigations were carried on regarding developments of a major project in the Vindhyan region in Uttar Pradesh, and in between, small easy developments were undertaken expeditiously. Rihand Dam was undertaken by the Government of Uttar Pradesh in the Vindyan region on a tributary of Sone. Activities were undertaken all over India. Some other major projects were Nagarjunsagar in Andhra Pradesh, Chambal Complex in Rajasthan and Madhya Pradesh, Harike in Punjab, Tungabhadra (an interstate project now shared by Karnataka and Andhra Pradesh), Bhadra and Ghatprabha in Karnataka, Lower Bhavani in Tamil Nadu, Matatila in Uttar Pradesh, and Mayrakshi in West Bengal. Several other major projects were undertaken in the Second Plan (1956–1961), such as Rajasthan Canal, the interstate Gandak Project in Bihar and Uttar Pradesh, Tawa in Madhya Pradesh, Kabini in Karnataka, Kansabati in West Bengal, Kadana, Ukai, and Bharuch (Narmada) in Gujarat, and Purna, Girna, Mula, and Khadakwasla in Maharashtra. More projects were undertaken in the Third Five-Year Plan (1961–1966) and the Three Annual Plans (1966–1969) when high dam developments in Himalayas were actively undertaken, now with a more scientific and knowledgeable background, indigenously.

An Irrigation Commission was appointed in 1969 "to go into the question of future irrigation development in the country." The report was submitted in 1972. The commission made an estimate of the utilizable water resources of the country. It was recommended that river basin plans should be developed. Domestic water was given the highest priority, followed by industry and then by irrigation. Between irrigation and power, the former was given priority. Closer collaboration was recommended between irrigation and agricultural departments, with irrigation demand to be determined by the latter. Conjunctive surface and groundwater development was recommended. It was also recommended that a portfolio of well-developed projects should be kept ready for speedy execution of appropriate projects. Soil conservation should be undertaken conjunctively with development of water resources, particularly in the more critical areas of the catchments. Cropping pattern should be undertaken commensurate with availability of water. Use of advanced techniques for field level irrigation should be increasingly adopted. Carry-over storage should be developed to improve on the current 75% availability criteria of planning. Command level development was given particular emphasis. Improvement to existing works was emphasized, as large numbers of them were very old. Storages were recommended to be developed urgently in view of the hydrologic–climatic conditions of the country. Considering that the scope for extension of agriculture had almost exhausted, a future increase in yield must be obtained from intensive and double cropping. Interbasin transfer as proposed at that time through the Ganga–Cauvery link was recommended to be investigated. Maintenance of ecological balance was recommended. Economic feasibility in terms of specified benefit/cost ratio of 1:5 was upheld, and it was further emphasized that the project should include supply of water up to farmer's field. The sector should be made financially sound. Recommendations about water rates were made. Administratively, setting up of River Basin Commissions was recommended. Setting up of a high-level authority, "The National Water Resources Council," to take policy decisions, with the Prime Minister as Chairman, was recommended. Union government was urged to take an active role in settling interstate disputes. Emphasis was laid on education and research. It should, however, be emphasized that all the recommendations were generalistic for the simple reason that the science of the subject had not yet developed in India, particularly in the profession. Paradoxically, the poor state of science in the profession continues, as the study of the recommendations of another recent National Commission on Integrated Water Resources Development (1999) brings out.

Development continued as before. Advances in water resources planning had taken place in terms of integrating engineering and economics backed by the science of decision making called systems planning. One of the activities of the Fourth Plan was to try to introduce this science among the professionals in the states through collaborative water resources planning, which was encouraged under the collaborative professional–academic studies under the author.

An important change was a dramatic increase in groundwater for irrigation. From a utilization figure of groundwater irrigation of 6.50 Mha in contrast to 9.70 Mha under surface water up to 1951, groundwater utilization jumped to 42.05 Mha in contrast to the figure of 28.20 Mha for surface irrigation in 1996–1997. Currently, groundwater accounts for 50% of the total irrigation. The development, however, has been quite uneven. Punjab, Haryana, Tamil Nadu, and Gujarat account for large groundwater development. It has been quite low in the eastern regions being about 8% in Orissa, 9% in Bihar, 20.5% in Assam, and 34% in Uttar Pradesh.

There has been an impressive development of irrigation in India, as shown in Table 3.2, with a developed potential of 90.82 Mha up to 1996–1997, which has been extended to 102.77 Mha up to March 2007, making India the global leader in area irrigated. The potential is estimated to be 139.89 Mha.

However, as we will discuss in Chapter 5, water resources development has remained focused on irrigation and that too with ad hocism. There has been little development in the most important activity—provision of drinking water. Irrigation has been undertaken with total disregard to environmental considerations. With the diversion of almost all low flows, the rivers have become open sewers, as urbanization has been taking place at an exponential rate even with a limited industrialization.

A serious predicament awaits us if the present trends continue.

TABLE 3.2

Cumulative Irrigation Development in the Plan Periods[a]

Item		Pre-Plan up to 1951	Up to Annual Plan 1996–1997	Ultimate Potential	Created Irrigation Potential as Percentage of Ultimate
(1) Major and	Potential	9.70	32.69	58.46	55.9
medium irrigation	utilization	9.70	28.20		
(2) Minor irrigation	Potential	6.40	12.25	17.38	70.5
(a) Surface water	utilization	6.40	10.82		
(b) Groundwater	Potential	6.50	45.88	64.05	71.6
	utilization	6.50	42.05		
Total of (2)	Potential	12.90	58.13	81.43	71.4
	utilization	12.90	52.87		
Grand total	Potential	22.60	90.82	139.89	64.9
(1) + (2)	utilization	22.60	81.07		

Source: National Commission for Integrated Water Resources Development (NCIWRD), Report, Vol. I, *Integrated Water Resources Development—A Plan for Action*, Ministry of Water Resources, Government of India, New Delhi, 1999.

[a] In million hectares.

4

Futures Challenges and Proposed Responses

4.1 Introduction

The challenges faced by the society in view of the management of water have led the government to appoint commissions from time to time. Water for Government meant irrigation, and the concerned Ministry was known as Ministry of Irrigation. The last Commission, an Irrigation Commission, was appointed in 1972 (GOI 1972). The Ministry of Irrigation, Government of India, was renamed Ministry of Water Resources, although different aspects of water continue to be looked after by other Ministries as before, and the Irrigation title continues in several States. The latest exercise is by the National Commission on Integrated Water Resources Development, appointed by the Government of India, which proposed a plan for action and is, therefore, referred to as NCIWRDP 1999.[1] It is presented in detail. Studies by some agencies dealing with the subject have also been made and are also briefly presented.

4.2 NCIWRDP 1999—Findings and Recommendations

NCIWRDP 1999 provides the current official perspective of water resources development and management of India. It is briefly summarized and reviewed first. Main findings and recommendations of the NCIWRDP 1999 are presented briefly as follows.[2]

The Commission noted in its opening statement, "While there have been Commissions on agriculture, irrigation and floods earlier, this is the first National Commission on Water Resources."

The Commission noted, at the end of the twentieth century, that the world faces a number of challenges affecting the availability, accessibility, use, and sustainability of its freshwater resources. Globally, the use of water has increased 35 times over the past three centuries. About 3250 km^3 of freshwater is withdrawn and used annually. Of this, 69% is used for agriculture, 2% for industry, and 8% for domestic use. Water use varies considerably around the world. In Africa, Asia, and South America, agriculture is the primary user. Asia uses 86% of its water for agriculture, mainly through irrigation, but in most of the European countries and North America, domestic and industrial requirements far exceed agricultural needs.

India, which has 16% of the world's population, has roughly 4% of the world's water resources and 2.45% of the world's land resources. The distribution of the water resources in the country is highly uneven over space and time. Over 80%–90% of the Indian river flows occur in four months of the year, and there are regions of harmful abundance and acute scarcity. Vast population lives in the latter area. While approaches, solutions, and actions are suggested to deal with the issues, they will be confronted more satisfactorily to the extent that we are able to develop a national consensus. Major attitudinal and organizational changes would be necessary.

4.2.1 Water Availability

The average annual precipitation in India including snowfall has been estimated as 4000 km^3. River basins are the natural hydrological units. The country has been considered in terms of 24 river basins. Annual mean flow is reckoned as the water resource of the basin. Region-wise breakup of the water resources has been estimated and is given in Table 4.1.

4.2.2 Utilizable Water Resources

Average flow and utilizable surface water, basin wise, were estimated by the NCIWRD 1999 and are given in Table 4.2. (Estimates by the author, which will be discussed later, are also given.) The average flow was estimated to be 1952.87 km^3, and the utilizable surface flows were estimated to be 690.31 km^3.

TABLE 4.1

Region-Wise Breakup of Water Availability

Sl. No.	Region	Water Availability		Percentage Area of Total (%)
		(km^3)	(%)	
1.	Ganga–Brahmaputra–Meghna	1200	60	33
2.	West-flowing rivers south of Tapi	200	11	3
3.	Remaining area	553	29	64
	Total	1953		

Source: National Commission for Integrated Water Resources Development (NCIWRD), Report, Vol. I, *Integrated Water Resources Development—A Plan for Action*, Ministry of Water Resources, Government of India, New Delhi, 1999.

4.2.3 Storages

For the use of surface water during the nonmonsoon months, there is a need for building up of storage capacities in reservoirs and storage tanks. The utilizable flows include water made available through storage projects. The total storage built up in the projects completed up to 1995 is about 174 km^3. From projects under construction, another 75 km^3 of storage capacity is likely to be added. Small tanks provide about 3 km^3. This gives a total available storage of 253 km^3. From identified future projects, another 132 km^3 can be added, making a total of 385 km^3. Basin-wise figures are shown in Table 4.3. It is important to emphasize that this is a bare minimum live storage needed to balance seasonal flows in an average year. Without the availability of this much storage, the assumption of 690 km^3 of utilizable surface flows will not be valid.

For lack of adequate storage sites on the Ganga in its catchment in India, the available flow cannot be fully used, and there is surplus flow in the river during monsoon. This problem is even more acute in Brahmaputra, where the available land is also very limited and the water cannot be used. Similarly, limited land is available in the west-flowing rivers of the western coastal areas.

Several features of water availability and utilizable potential need to be emphasized. First, there is a vast difference between annual surface water flows and their potential for utilization in view of the hydrological and physiographic characteristics. About 80%–90% of the precipitation is in four monsoon months and that too in few days. Second, the spatial and temporal distribution of rainfall is so uneven that the annual averages have very little significance for all practical purposes. For example, there is acute scarcity of water in the southern and western tributaries of Ganga, whereas there is excess in the eastern parts. Similarly, the Godavari basin becomes water rich only after the confluence of Pranhita—the upstream region is water scarce. In fact, one-third of the country is always under threat of drought not necessarily due to deficient rainfall but many times due to its uneven occurrence. This peculiar phenomenon of "scarcity under plenty" often manifests itself in the form of droughts and floods at the same time even in the same river basin.

4.2.4 Utilizable Groundwater Resources

In planning development of water, only dynamic freshwater resources have been considered by the governmental organizations as well as the NCIWRD 1999. Total replenishable groundwater is estimated at 432 km^3. Out of this, 396 km^3 is considered replenishable. Nearly 50% of the irrigation in the country is by groundwater. Groundwater also occurs in aquifer zones below the zone of water level fluctuation, called static groundwater.

Central Ground Water Board (CGWB) had prepared a National Perspective Plan (1996) for recharge of groundwater by utilizing surplus monsoon runoff in river basins. There is also the possibility of availability of artesian

TABLE 4.2

Water Resources Availability and Utilization[a]

Sl. No.	River Basin	Surface Water			Groundwater			Total Water	
		Mean	Utilizable Commission	Chaturvedi	Replenishable Commission	Artificial Recharge Chaturvedi	Total Chaturvedi	Commission	Chaturvedi
1	Indus	73.31	46.0	46.0	14.29	20.7	34.99	60.29	80.99
2	Ganga–Brahmaputra–Meghna basin								
2a	Ganga sub-basin	525.02	250.0	250.0	136.47	94.9	231.37	386.47	481.37
2b	Brahmaputra	629.05	24.0	24.0	25.72	15.9	41.62	49.72	65.62
2c	Meghna	48.36	–	–	8.52	–	8.52	8.52	8.62
3	Subarnarekha	12.37	6.81	6.81	1.68	0.6	2.28	8.49	9.09
4	Brahmani–Baitarani	28.48	18.3	18.30	3.35	0.4	3.75	21.65	22.05
5	Mahanadi	66.88	49.99	66.88	13.64	1.5	15.14	63.63	82.02
6	Godavari	110.54	76.30	110.54	33.48	5.8	39.28	109.78	149.82
7	Krishna	69.81	58.00	69.81	19.88	3.3	23.18	77.88	92.99
8	Pennar	6.32	6.86	6.86	4.04	–	4.04	10.90	10.99
9	Cauvery	21.36	19.00	21.36	8.79	1.0	9.79	27.79	31.15
10	Tapi	14.88	14.50	14.88	6.67	–	6.67	21.17	21.42
11	Narmada	45.64	34.50	45.64	9.38	1.4	10.78	43.88	56.92

12	Mahi	11.02	3.10	11.02	3.5	1.4	4.9	6.60	15.92
13	Sabarmati	3.81	1.93	3.81	2.9	0.9	3.8	4.83	7.61
14	West-flowing rivers of Kachchh, Saurashtra, and Luni	15.1	14.98	14.98	9.1	3.3	12.4	24.08	27.38
15	West-flowing rivers south of Tapi	200.94	36.21	36.21	15.55	2.1	17.65	51.76	53.86
16	East-flowing rivers between Mahanadi and Pennar	22.52	13.11	13.11	12.82	0.6	13.52	25.93	26.63
17	East-flowing rivers between Pennar and south of Cauvery	16.46	16.73	16.73	12.65	4.2	16.85	29.38	33.58
18	Rivers draining into Bangladesh	8.57	NA	NA	–	–	–	–	–
19	Rivers draining into Myanmar	22.43	NA	NA	–	–	–	–	–
	Total	1952.87	690.32	776.94	342.43	158.00	500.43	1032.75	1276.47

Source: Commission figures from National Commission for Integrated Water Resources Development (NCIWRD), Report, Vol. I, *Integrated Water Resources Development—A Plan for Action*, Ministry of Water Resources, Government of India, New Delhi, 1999. Chaturvedi figures are from Chaturvedi, M.C., *Water Policy* 3, 297–320, 2001.

Note: In cubic kilometers.

[a] In cubic kilometers.

TABLE 4.3

Storages in India—Basin-Wise

		Live Storage Capacity (km^3)			
Sl. No.	River Basin	Completed Projects	Projects under Construction	Total	Projects under Consideration
1	2	3	4	5	6
1	Indus	13.83	2.45	16.28	0.27
2	Ganga–Brahmaputra–Meghna basin				
2a	Ganga subbasin	36.84	17.12	53.96	29.56
2b,c	Brahmaputra subbasin	1.09	2.40	3.49	63.35
3	Subarnarekha	0.66	1.65	2.31	1.59
4	Brahmani–Baitarani	4.76	0.24	5	8.72
5	Mahanadi	8.49	5.39	13.88	10.96
6	Godavari	19.51	10.65	30.16	8.28
7	Krishna	34.48	7.78	42.26	0.13
8	Pennar	0.38	2.13	2.51	NA
9	Cauvery	7.43	0.39	7.82	0.34
10	Tapi	8.53	1.01	9.54	1.99
11	Narmada	6.6	16.72	23.32	0.47
12	Mahi	4.75	0.36	5.11	0.02
13	Sabarmati	1.35	0.12	1.47	0.09
14	West-flowing rivers of Kachchh, Saurashtra, and Luni	4.31	0.58	4.89	3.15
15	West-flowing rivers south of Tapi	17.34	4.97	22.31	2.54
16	East-flowing rivers between Mahanadi and Godavari	1.63	1.45	3.08	0.86
17	East-flowing rivers between Godavari and Krishna				
18	East-flowing rivers between Krishna and Pennar				
19,20	East-flowing rivers between Pennar and Cauvery and south of Cauvery	1.42	0.02	1.44	NA
21	Area of North Ladakh not draining into Indus	NA	NA	NA	NA

(continued)

TABLE 4.3 (Continued)

Storages in India—Basin Wise

		Live Storage Capacity (km³)			
Sl. No.	River Basin	Completed Projects	Projects under Construction	Total	Projects under Consideration
1	2	3	4	5	6
22,23	Rivers draining into Bangladesh and Myanmar	0.31	0	0.31	NA
24	Drainage areas of Andaman, Nicobar and Lakshadweep Islands	NA	NA	NA	NA
	Total	173.71	75.43	249.14	132.32
	Say	174	76	250	132

Source: National Commission for Integrated Water Resources Development (NCIWRD), Report, Vol. I, *Integrated Water Resources Development—A Plan for Action*, Ministry of Water Resources, Government of India, New Delhi, 1999.

groundwater in the Ganga–Brahmaputra–Meghna (GBM) basin. The possibility of additional water through these technological options is currently not being considered by the governmental planning agencies and has not been considered by NCIWRD 1999.

4.2.5 Total Utilizable Water Resources

Total utilizable water resources of the country are, thus, estimated to be 690 km³ of surface water and 396 km³ of groundwater, totaling to 1086 km³.

4.2.6 Floods

About 40 million hectares (Mha) is prone to floods, though not all the vulnerable areas get affected each year. On the other hand, about one-third of the country is afflicted by recurring droughts.

4.2.7 Water Requirement

Requirement is closely related to population, demand for food, production of non-food agricultural and industrial items, production of energy and improvement in the quality of life, and preservation of ecology and environment. Estimates of water requirements have been made for the years 2010, 2025, and 2050. Population for the year 2050 has been estimated as

1581 million (higher limit) and 1346 million (lower limit). Several assumptions have to be made. The goal of food self-sufficiency at the macrolevel is one.

Total water requirements in the country were estimated as 694–710, 784–850, and 973–1180 km³ by the years 2010, 2025, and 2050, respectively, depending on the high demand and low demand scenarios. Irrigation would continue to have the highest water requirement, between 628 and 807 km³ (or about 68% of the water requirement), followed by domestic water requirements, at about 90–111 km³ (or about 10% of the water requirements), in the year 2050. The projected use per capita in the year 2050 would be about 725–750 km³ as compared to about 650 km³ at present.

The country's total water requirements in the year 2050 barely match the estimated utilizable water resources. It is of paramount importance that we should aim at reducing water requirement to the low demand scenario. Some major considerations appear. First, utmost efficiency should be introduced in water use. Second, average availability at national level does not imply that all basins are capable of meeting their full requirement from internal resources. Third, the issue of equity in the access to water, between regions and between sections of population, assumes greater importance in what is foreseen as a fragile balance between regions and between the aggregate availability of water.

4.2.8 Irrigation, Flood Management, Hydropower, and Navigation

Irrigation has been important in India from historical times. Since independence, irrigated area rose from 22.60 to 80.76 Mha up to 1997 against the ultimate target of 140 Mha, of which groundwater has been nearly 50%. Such a large program has been implemented entirely with indigenous technology and equipment.

Some lessons that emerge are that projects should be undertaken only after they are properly planned and funding is assured; otherwise, there are delays in construction and cost overruns, as in the case of most of the projects. There has been a disproportionately large lag in potential created and utilized. Modernization of irrigation is overdue.

Waterlogged areas need to be reclaimed. Intrusion of salinity in coastal areas and problems arising from pollution from industrial effluents have to be rectified. Much greater effort and attention need to be given to management aspects of irrigated agriculture both from the technical point of view and to the involvement of the users.

Focusing on command area development programs, there is a great scope for upgrading management practices and improving water use efficiency. The reform agenda would include technological improvements, institutional reforms, and financing of irrigation schemes including private sector participation.

About 40 Mha is prone to floods, though not all the vulnerable areas get affected each year. On the other hand, about one-third of the country is afflicted by recurring droughts. Of the flood-prone area of 40 Mha in the country, reasonable protection has been provided to about one-third of the area by now. Complete protection is not possible, and strategy should aim at efficient management of floods, flood proofing including disaster preparedness and response planning, flood forecasting and warning, and other nonstructural measures such as disaster relief, flood fighting including protection of public health, and introduction of flood insurance.

Hydropower is one of the important uses of water and its advantages are well known. However, in spite of the large hydropower potential of 84,000 MW at 60% load factor, the present development has been only to the tune of 13,000 MW. In addition, 56 sites were identified for Pumped Storage Schemes with a total installation capacity of 96,000 MW. Only 1554 MW had been developed, 4340 MW was under construction, and only one, Tehri St.-II with a capacity of 1000 MW, in Ganga basin, had been approved by Central Electricity Authority. (The original design of Tehri St.-II did not envisage pumped storage. The change was made by the author, who was on the Board of Consultants.)

Navigation is important. Inland navigation should be encouraged by treating it as nascent industry for some years.

4.2.9 Domestic Requirement

Domestic water supply is most important. Almost 100% coverage in both water supply and sanitation was recommended as early as 1949, but we are nowhere near it. There are large shortfalls and failings. The sector needs utmost attention.

4.2.10 Water for Industrial Uses

The sector also shows vital failings and needs improvement.

4.2.11 Local Water Resources Development and Management

Integrated watershed development is important and requires much improvement. A major reformulation of priorities and programs and restructuring of institutions and operational means are vital for integrated local watershed development. Recommendations in this connection were made by the Commission.

4.2.12 Interbasin Transfers

In view of large disparities of the availability of water in different basins, interbasin transfer of water has been receiving attention. The overall approach is that economic development of no part of the country should be constrained by shortage of water; at the same time, the pattern of development could be different in States having adequate water and those to which water may have to be transferred at high cost.

The National Perspective brought out by the Ministry of Water Resources has two main components: (1) Himalayan Rivers Development and (2) Peninsular Rivers Development. The Himalayan Rivers Development envisages construction of storages on the main Ganga and the Brahmaputra and their principal tributaries in India and Nepal along with interlinking to transfer surplus flows of the eastern tributaries of Ganga to the west. Peninsular component envisages interlinking of Mahanadi–Godavari–Krishna–Cauvery Rivers, interlinking of west-flowing rivers north of Bombay and south of Tapi, interlinking of Ken–Chambal, and diversion of other west-flowing rivers.

The Himalayan component data are not freely available, but based on published information, it appears that this component may not be feasible by 2050. The peninsular component of the proposals is briefly described as follows. Nine links are proposed for interlinking east-flowing peninsular rivers. These links involve construction of five dams and nine link canals. The headworks will submerge 2.5 lakh ha and require rehabilitation of more than 4.5 lakh persons (1 lakh = 100,000). The links will also require agreements among all the concerned States, and such concurrence would be facilitated by some form of *quid pro quo*.

The studies indicate that based on mean annual flows except for Krishna (if irrigation intensity is adopted at rather high 45%), Cauvery, and Vaigal, the balance is positive in other cases. The shortage in Cauvery is 12% of gross demand and that in Vaigal is 16%. These shortages result after increasing the present irrigated area to 1.4 times in case of Cauvery and 1.6 times in case of Vaigal, and assuming the return flow at 60% of the imbalance. In case the return flow is taken as 80% of the imbalance, there is no shortage in Krishna, and those in Cauvery and Vaigal are reduced to 5% and 8%, respectively. Thus, there seems no imperative necessity for massive water transfer. The assessed needs of the basins could be met from full development and efficient utilization of intrabasin resources except in the case of Cauvery and Vaigal basins. Some water transfer from Godavari toward the south should take care of the deficit in Cauvery and Vaigal basins. This could be accomplished by constructing a low dam at Inchampali involving minimal submergence.

4.2.13 Financial Aspects

The financial aspects of the water resources sector are a matter of very grave and urgent concern. Public investment, as a percentage of overall plan outlays, has gone down from 22.5% in the initial plans to 6.5%. Some projects started in 1950 are still not completed. Cost and time overruns have become endemic, and spillover costs are increasing with every plan. Financial returns are negligible as a result of highly subsidized pricing of water and substandard modernization. The returns are not even adequate to cover costs of operation and maintenance. Maintenance has been neglected and systems are deteriorating. There is little regard in the entire system for prudential norms and accountability. The situation continues to deteriorate over the years. Unless urgent remedial measures are taken to reverse the trend, the economy and the people will have to pay a dear price. The whole question is examined based on this assessment.

While public outlays have to be enhanced and better applied based on priorities, it is necessary to augment the available resources through institutional finance, water equity and bonds, private sector participation, commercial exploitation of land, and people's participation. For field level works, in addition to rural area programs, the community's involvement and participation could be significant.

A number of measures are suggested for immediate adoption to instill a financial discipline in the system. These include project-wise funding, proper preparation of cost estimates including possible escalation, timely revision of such estimates, clear definition of commencement and completion of projects, insistence of completion reports, and limitation on the change of scope of a project and on establishment costs.

Water used for irrigation is an economic good, and its logical pricing is a key to improving water allocation and encouraging conservation. The present rates need substantial revision. The water rates should cover the entire annual operations and management cost plus 1% of the gross value of the cereal and a higher percentage in respect of others. The objective should be toward volumetric measurement through stages.

4.2.14 Legal Issues

There are a number of complex legal issues when matters regarding integrated development of interstate rivers, allocation of river waters, interbasin transfers, and utilization of groundwaters, water rights, and people's participation are considered. Valuable suggestions were given.

4.2.15 Institutional Aspects

Institutional reorganization at all levels is urgently required. Water has diverse uses and interdisciplinary management is essential. The subject has been considered in detail.

4.2.16 Project Planning and Prioritization

There is a need to make a wide range of changes in approaches to project planning, particularly with regard to allocation of water among various uses, dependability and carry-over related issues, conjunctive use, project size, and viability area. The present procedures of cost–benefit analysis are too simplistic and reutilized.

Project implementation requires considerable improvement. There is an urgent need for prioritization of schemes.

4.2.17 International Dimensions

The largest part of India's water resources is contributed by Transboundary Rivers, which rise from and flow into or through six of its neighbors, namely, Nepal, Bhutan, Tibet/China, Myanmar, Bangladesh, and Pakistan. Taking cognizance of this geopolitical reality, it is obvious that no meaningful or optimal national water resource planning is possible in relation to these rivers without bilateral or regional collaboration.

The problems and type of issues that usually arise in working out such cooperation as well as those that have emerged or are emerging as a result of global concern for environment, biodiversity, emission, resettlement, and rehabilitation of displaced persons are indicated. The subject needs much consideration as to how it could be obtained. Detailed studies of development would contribute much to facilitating collaborative development.

4.2.18 Water Quality and Environmental Aspects

Along with quantitative water availability, it is also important that water quality is appropriate. The source of all water supplies is rainwater, and even this source is polluted, even in normal course, from natural processes. With increasing economic activities, it is further polluted. As water follows the hydrological cycle, interaction with land and fauna and flora takes place and pollution increases, which is further compounded on account of human and economic activities.

The Central Pollution Control Board in association with State Pollution Control Boards has established a water quality monitoring network with 480 stations spread over 21 states and 4 union territories in the country. Currently, 25 water quality parameters are measured periodically. All the rivers are severely polluted. The groundwater is also getting increasingly polluted.

The challenge in water resources development is to balance the needs of development and those of environmental health and thereby ensure the sustainability of development. Integrated Water Resources Management (IWRM) should be based on the perception of water as an integrated part of the ecosystem and as a natural resource whose quantity and quality determine the nature of its utilization.

4.2.19 Research and Development Needs

Water resources development is to be seen not merely as a single-sector-end objective, but as a prime mover in developing larger systems with multiple linkages. This calls out for a well set out multidisciplinary research agenda covering not only technological issues but also issues of social, economic, legal, and environmental concerns. With a trained, motivated manpower being the backbone of any developmental activity, also in the water resources sector, there is a need for human resources development. The challenge in the water sector is to simultaneously take care of the needs of development and environmental health, and thereby ensure the sustainability of development. The problems are not beyond the present state of knowledge and technology. Given the needed will and societal awareness, the nation shall be able to meet the challenge.

4.3 NCIWRDP 1999—A Review

The Commission made meaningful recommendations on various aspects of water resources development, which can be summarized as urgent necessity to modernize the sector, but they were more of pious pronouncements than leading to any policy setting. In several important respects, it was blind. For instance, irrigation is a dominant user of water, accounting for about 83%. Yet it is extremely inefficient in terms of use of water and quality of service provided to agriculture. The field level wastes are high. Irrigating the largest area in the world is one of the worst performers as measured by agricultural output (World Resources Institute 1996). In terms of future targets, it set an achievement of yield by India in 2050 of 4000 MW. The issue of interlinking of rivers continued to obsess it.

The Commission was, essentially, a continuation of the current Governmental system, and the entire activity was an extension of the current Governmental thinking. Indeed, this is a matter of greatest concern and is discussed in Chapter 7.

4.4 Estimate of Utilizable Water Resources and Demand—Water Balance

NCIWRD 1999 adopted the estimates of the utilizable surface waters as developed by Central Water Commission, with some modifications. The basis of estimated flows considered utilizable is not available because it is classified. It has, however, been stated that these have been worked out based on estimates of using them by diversion and storing the surface flows.

Some revolutionary technologies will be discussed in Chapter 5, which will demonstrate that the utilizable potential in the GBM is much more than has been currently estimated by the NCIWRD, if we get out of the historic trap of the current technology based on "using them by diversion and storing the surface flows." However, we have not proposed any change in estimated water availability of the GBM Rivers at this stage of the following discussion, except that the utilizable potential in peninsular rivers should be up to mean flows as there is no problem about carry-over storage. Our estimates in Table 4.2 are based on this.

Regarding groundwater, the CGWB has estimated that 342.4 km³ could be developed on a sustainable basis. NCIWRD 1999 has adopted this figure for groundwater potential. CGWB has, however, also stated that, in addition to the 342.4 km³, 214 km³ of surplus monsoon runoff could be stored in groundwater through conventional techniques, of which 158.0 km³ is considered retrievable on a sustainable basis. This figure has not been adopted by the NCIWRD on the plea that these need to be firmed up by detailed study.

Our view is different. The subject has been studied in detail.[3] Scientific analysis, already conducted, confirms that the technology proposed by CGWB is feasible. We will demonstrate in Chapter 5 that much higher groundwater recharge is possible through certain novel technologies. Therefore, we have adopted the recommended recharge by the CGWB and have adopted the revised figure of 342.43 km³ plus 158.00 km³, amounting to 500.43 km³. The figures, as developed by the NCIWRD 1999 and as proposed by the author, limiting to the technology currently used, are given in Table 4.2 in terms of each river basin.

It will be seen that, even after keeping ourselves confined to the current conventional technology, the surface water potential can be increased from 690.32 to 776.94 km³, the groundwater potential can be increased from 342.43 to 500.43 km³, and the total water availability is increased from a currently estimated value of 1032.75 to 1276.47 km³, an increase of about 12.5%, 45%, and 24%, respectively, even through conventional technology. The utilizable figures of water can be increased substantially if some revolutionary technologies are adopted as discussed in Chapter 5. As we will elaborate and emphasize later, the issue is not of figures but concepts and attitudes, first and foremost.

4.5 Water Requirement

NCIWRD 1999 carried out an estimate of demands up to 2050. They are based on certain assumptions, which in our judgment are seriously restrictive. The salient points are discussed as follows. The estimate of the Commission is given in Table 4.2. The proposed modifications are discussed as follows.

4.6 Review of the Official Demand Estimates and Proposed Modifications

The estimation of future demand determines the perspective of future development of the water resources and economic sector. Revolutionary change can be and have to be introduced both from economic and sustainability considerations. For example, the economy will increase by an order of magnitude, putting severe strain on the environment, including water. Formidable challenges on account of the climatic change will take place. These were not addressed by the Commission and are discussed in Chapter 5.

Essentially, the commission carried out an extrapolation of the current system. It did not analyze the shortcomings, much less make any positive suggestions. For example, the Commission did not even point out that our yields are one of the lowest in the world. The averages obtained in some of the large cereal-producing countries are given in Table 4.4.

It may be noted that China has already more than double the yield of India. The Commission, with all its wisdom, set a figure of 4000 kg/ha to be achieved in 2050, which has already been surpassed in China and is about two-thirds of the yield already achieved in the industrialized countries. However, in the first instance, we may examine the figures adopted by the NCIWRD 1999, as it represents the official perspective.

4.6.1 Agricultural Sector

The agricultural sector will continue to have considerable importance in the economy for a long time to come. Increasingly and rapidly, the agricultural sector has to become the most modern, as it will not be possible to ensure livelihood, food security, and environmental sustainability if the Third World agriculture continues. It has important implications regarding water demand and its service.

TABLE 4.4

Current Cereal Yield in Some Countries

			Water Use		Average Production
Sl. No.	Country	Yield (kg/ha)	Withdrawal (%)	Agriculture (%)	of Cereal (100 metric tons)
1	Egypt	5921	97	85	14,722
2	China	4482	16	87	401,088
3	India	2062	18	93	206,608
4	USA	5092	19	42	323,029
5	Japan	5588	17	50	13,603
6	Korea	5815	42	46	7523
7	France	6517	19	15	56,637
8	Germany	5588	27	20	35,568
9	Italy	4739	34	59	19,500
10	United Kingdom	6609	17	3	20,399

Source: World Resources Institute, *World Resources 1996–97*, Oxford University Press, New York, 1996.

Therefore, urgent modernization of the entire spectrum of activities is required on a well-planned and organized basis as an interdisciplinary interorganizational project. It is not an issue of appointing a National Commission for some activity, for example, irrigation, at one time, which makes *ad hoc* recommendations and then letting the old colonial processes continue. Appropriate land use policy, cropping intensity, irrigation intensity, reduction in delta, and achievements of appropriate yields are vital matters that will have important bearing on water demand, and all of these have to be managed.

Focusing on yield, it is seen that even without considering the modern advances in the field of biotechnology, much higher yield than that adopted by NCIWRD 1999 can be and must be achieved at the earliest. It is considered that adoption of a yield figure of 6000 kg/ha for cereals by 2050, which has already been achieved in most of the industrialized countries, in contrast to the currently recommended figure of 4000 kg/ha by the NCIWRD 1999, is logically imperative. Much higher figures will be justified (Conway 1998). Implicit in the specification of the proposed figure is the modernization of agriculture over the entire sector. Assured, adequate, and timely water supplies are critical determinants. Water has to be applied with increasing efficiency in terms of productivity and environmental conservation. The issue is not just of providing an apology of irrigation on a highly subsidized basis for Third World subsistence agriculture, as the current scene shows. The central issue is embedding development of water with modernization of agriculture for rapid economic development and environmental conservation. Thus, for a preliminary exercise, at least a figure of 6000 kg by 2050 AD should be adopted. All it means is that the issue is assuring quality of irrigation to provide adequate, timely, and reliable yields as part of the agricultural system modernization. In addition, development of increasingly scientific methods of water use in agriculture is the topmost priority in water management. This will lead to considerably reduced delta and preservation of water. However, in the estimation of demand, we have continued to adopt the current delta in the first instance.

We have developed two strategies for estimating water demand. In Strategy 1, all other assumptions are maintained the same as those of the NCIWRD except that the yield of the irrigated area by 2050 is specified to be 6.0 tonnes against the NCIWRD figure of 4.0 tonnes. The water demand drops by 29%.

An arbitrary assumption is usually made about the share of groundwater and surface water in providing irrigation in estimating future water resources development (NCIWRD). In Strategy 2, we have dropped that assumption and put primacy on groundwater with an arbitrary limit of allocating 400 km^3 against the estimated potential of 500 km^3. (This should be further increased, but tentatively, a low figure has been adopted.) Groundwater can provide much more reliable supplies than public surface water supplies. The groundwater requirement is considerably less than that of the surface water, the respective deltas being 0.49 and 0.61, and contributes to higher productivity. All other performance characteristics are maintained the same as before, and even then, only 56% of conventional water requirements are needed.

4.6.2 Domestic Water Demand

Domestic demand is the first priority. The total quantity is comparatively small. With increasing socioeconomic development, urbanization and industrialization are going to take place rapidly. From an environmental systems perspective, an issue of utmost importance is to ensure appropriate development of these sectors, as water supplies and quality cannot be assured unless proper habitat development, with appropriate spatial distribution, takes place. Thus, the issue is not merely that of adopting unit figures in these sectors but ensuring appropriate sectoral development.

Regarding urban water supplies, in our view, considering the future perspective of increasing socioeconomic development and the hot climate of the country, and the fact that the infrastructure system has to be developed in a long perspective, reviewing the global water developments, a figure of 400 liters per person per day (lpcd) appears to be reasonable. This may be contrasted with the figure of about 110–220 lpcd currently proposed and even more lowly figures recommended by some in terms of meeting the basic needs.[6] Conjunctively, in view of the extreme scarcity of water, the highest quality of return flows will also have to be specified. The important issue is not of the estimated figure but the fact that unless the highest emphasis is accorded to habitat and infrastructure development, it will not be possible to achieve even the minimal environmental specifications for water. Moreover, obtaining proper domestic water supplies will require that reallocation be made from the agricultural sector to domestic and environmental sectors. Thus, appropriate regional planning is extremely important if domestic and environmental requirements are to be met.

Considering rural water supply, resource scarcity is not an issue as it can be met by watershed management and rainwater harvesting. The issue is about mobilizing people and guiding them. For example, considering the GBM basin, people are sitting on an ocean of freshwater, which can be had just by digging some wells and lifting water. Even in the drought-prone areas, piped water supply can be provided as the first option. People's attitudes and views were investigated and are given in Appendix 2. The subject, therefore, is related to the wider issue of rural development.

4.6.3 Energy

The energy sector is going to develop rapidly and shall have the second largest demand for water. There is also a close interlinkage in terms of generation in view of the important role of hydroelectricity. Thus, water resources developmental planning has to be undertaken integrally with energy systems development. It is not a simplistic issue of hydrothermal balance, as generally emphasized currently.

4.6.4 Industrial Demand

Water requirements of industries have been studied in detail. There is considerable variation in unit figures depending on how conscious the society is regarding environmental management. Emphasis is now being laid on the subject in terms of the concepts of industrial ecology. The subject has not been given due emphasis in water resources management, but is of great importance in view of the serious environmental challenges due to be faced in view of serious problems of pollution, including toxic wastes, as industrialization takes place, and institutional arrangements will only develop with a certain inertia. We have, however, adopted the officially estimated figures (NCIWRD 1999) for the sector water demand, just for the sake of working out the overall water balance and estimating the minimum surface flows from environmental considerations. The highest priority has to be given to managing return flows in terms of quality and quantity via industrial ecology.

4.6.5 Environmental Demands

Minimum flow and appropriate water quality specifications from environmental considerations have often been neglected, particularly in the developing countries. Specific guidelines are not available as yet, but the issue is of great importance. Research is required on the subject, but the figure of 10%–15% of mean annual flow, as sometimes specified, is not correct, as this is arbitrary and does not specify the lean flow period requirements, which are the dominant determinants. Furthermore, the figure has to be related to the return flow figures. However, we have continued with the use of official figures, just for the sake of illustration. Besides specifications for the terrestrial waters, consideration of water quality at the water's edge is conjunctively related to the subject.

4.6.6 Water Demand Estimates

The figures of water withdrawal and return flow based on the Commission estimates and the proposed environmental systems concepts (under the caption Chaturvedi) are shown in Table 4.5. A significant decrease can be observed from the table. The demand estimates by Chaturvedi are still conservative, as potential for decreasing demand by adopting better irrigation practices, which is extremely important from economic and environmental considerations, has not been taken into account.

Based on the above-mentioned considerations, the total demand drops from 1180 to 950 and 826 km^3, a change of about 19% and 30%, respectively. It will be seen that a major contribution toward achieving sustainability is made. These

TABLE 4.5

Water Withdrawal and Return Flow Projections

(a) Withdrawal (km³)

Sl. No.	Use	Year 1997–1998		Commission 2050			Chaturvedi 2050			
		Withdrawal	%	Low	High	%	Low	High Strategy 1	High Strategy 2	% Strategy 2
1	Irrigation	524	83	628	807	68	407	577	453	55
2	Domestic	30	5	90	111	9	90	111	111	13
3	Industries	30	5	81	81	7	81	81	81	10
4	Power	9	1	63	70	6	63	70	70	8
5	Inland navigation	0	0	15	15	1	15	15	15	2
6	Flood control	0	0	0	0	0	0	0	0	0
7	Ecology	0	0	20	20	2	20	20	20	2
8	Evaporation loss	36	0	76	76	7	76	76	76	10
9	Total	629	100	973	1180	100	752	950	826	100

(b) Return Flows—Year 2050 (km³)

Sl. No.		Irrigation		Domestic		Industries		Total	
		Low	High	Low	High	Low	High	Low	High
1	Groundwater	115	147	7	8	–	–	122	155
2	Surface water	13	17	36	47	40	40	91	104
3	Total	128	164	45	55	40	40	213	259

Note: Return flow figures are from NCIWRD 1999. Chaturvedi figures are from Chaturvedi (2001).

TABLE 4.6

Estimates of Water Availability (Regional)

	Region	Commission			Chaturvedi		
		Surface	Ground	Total	Surface	Ground	Total
1.	Himalayan region	320.00	185.00	505.00	320.00	316.50	630.50
2.	Peninsular region	264.18	102.28	388.46	350.80	117.56	468.36
3.	Rest	106.14	55.15	161.14	106.14	66.47	171.61
4.	Total	690.32	324.43	1032.75	776.94	500.53	1276.47

estimates are, however, only indicative because the position varies considerably spatially, and each river basin will have to be studied in detail. The advances will be even more if agricultural and other sectors are managed scientifically.

Estimates of water availability as adopted by the Commission and modified by us, based on the suggestions we made as discussed earlier, are given in Table 4.6.

A perspective of development according to the official GOI perspectives is shown in Figure 4.1. The perspective, even with the modifications as

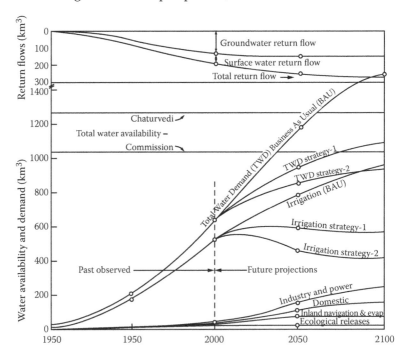

FIGURE 4.1

Development of India's waters—actual and future perspective. (From Chaturvedi, M.C., *Water Policy* 3, 297–320, 2011.)

proposed, confining ourselves to the current technologies, is also shown in Figure 4.1. It is seen that the issue is not that of looming unavailability of water, although considerable strain will be experienced, but the urgent need for modernization of the sector. The picture is even better transformed if the novel technologies and the environmental systems approach discussed in the following chapters are adopted. The issue of the currently proposed interlinking of India's rivers drops dead. It is, however, discussed a bit in detail in the next section, as it continues to be an important part of the official thinking.

4.7 Interlinking of India's Rivers

4.7.1 Introduction

The former Ministry of Irrigation (now Ministry of Water Resources) and the Central Water Commission, Government of India, formulated a National Perspective Plan for optimum utilization of water resources in the country. It envisaged interbasin transfer of water from surplus to deficit areas, through interlinking of rivers of India. It was called the Inter Linking of Rivers (ILR) proposal. A National Water Development Agency was established by the Government of India to examine and formulate the scheme further. If implemented, it will be one of the world's largest and most costly engineering activities. It will have far-reaching socioeconomic–environmental impacts.

It is presented a bit in detail because, even though it was demonstrated by the Working Group established by the NCIWRD 1999 that it is unscientific and is not required, it continues to be pursued with vigor. Even the Supreme Court is stated to have upheld it.

4.7.2 National Perspective Plan and ILR Proposal

The proposed National Perspective Plan comprises two main components: (1) Peninsular Rivers Development and (2) Himalayan Rivers Development. These are shown in Figures 4.2 and 4.3, respectively.

(1) Peninsular Rivers Development

The scheme is divided into four major parts:

(a) Interlinking of Mahanadi–Godavari–Krishna–Cauvery Rivers and building storages at potential sites in these basins
(b) Interlinking of west-flowing rivers, north of Bombay and south of Tapi
(c) Interlinking of Ken–Chambal Rivers

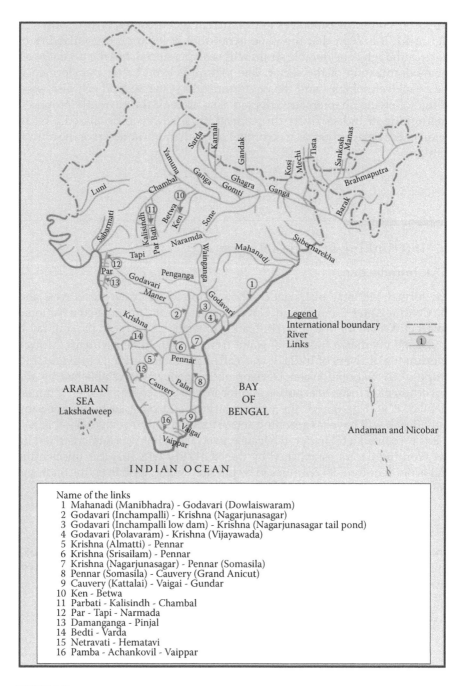

Name of the links
 1 Mahanadi (Manibhadra) - Godavari (Dowlaiswaram)
 2 Godavari (Inchampalli) - Krishna (Nagarjunasagar)
 3 Godavari (Inchampalli low dam) - Krishna (Nagarjunasagar tail pond)
 4 Godavari (Polavaram) - Krishna (Vijayawada)
 5 Krishna (Almatti) - Pennar
 6 Krishna (Srisailam) - Pennar
 7 Krishna (Nagarjunasagar) - Pennar (Somasila)
 8 Pennar (Somasila) - Cauvery (Grand Anicut)
 9 Cauvery (Kattalai) - Vaigai - Gundar
10 Ken - Betwa
11 Parbati - Kalisindh - Chambal
12 Par - Tapi - Narmada
13 Damanganga - Pinjal
14 Bedti - Varda
15 Netravati - Hematavi
16 Pamba - Achankovil - Vaippar

FIGURE 4.2
Peninsular Rivers Development component. (From National Commission for Integrated Water Resources Development (NCIWRD), Report, Vol. I, *Integrated Water Resources Development—A Plan for Action*, Ministry of Water Resources, Government of India, New Delhi, 1999.)

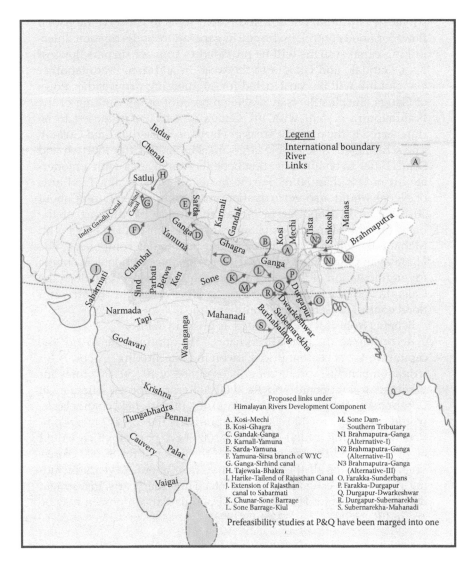

FIGURE 4.3

Himalayan Rivers Development component. (From National Commission for Integrated Water Resources Development (NCIWRD), Report, Vol. I, *Integrated Water Resources Development—A Plan for Action*, Ministry of Water Resources, Government of India, New Delhi, 1999.)

(d) Diversion of other west-flowing rivers toward the east

Of the four, the first is the most important and will be the focus in the following discussion while considering the peninsular component.

(2) Himalayan Rivers Development

Himalayan Rivers Development envisages construction of storages on the main Ganga and the Brahmaputra Rivers and their principal tributaries in India, Bhutan, and Nepal so as to conserve monsoon flows for flood control, hydropower generation, and irrigation. Interlinking canal systems will be provided to transfer surplus flows of Kosi, Gandak, and Ghagra to the west. In addition, Brahmaputra–Ganga Link will be constructed for augmenting dry weather flows of Ganga. Surplus flows available on account of interlinking of the Brahmaputra system with the Ganga system are proposed to be transferred to the drought areas of Haryana, Rajasthan, and Gujarat. The scheme will also enable large areas in South Uttar Pradesh and South Bihar to obtain irrigation benefits from Ganga with a moderate lift of less than 30 m. Furthermore, all lands in the Terai area of Nepal would also get irrigation apart from generation of about 30 MkW of hydropower in Nepal and India. Interlinking of the main Brahmaputra and its tributaries with Ganga and Ganga with Mahanadi will enable transfer to the peninsular component. The Himalayan Component would also provide the additional discharge for augmentation of flows at Farakka required *inter alia* to flush the Calcutta Port, the inland navigation facilities across the country, and flood control in Ganga and Brahmaputra.

Benefits from ILR are estimated as given in Table 4.7.

According to the official estimates, "the approximate cost of the capital works, at current prices, involving interlinking proposals for irrigation benefits, can be roughly assumed at Rs. 50,000 crores for peninsular component and Rs. 1,00,000 crores for the Himalayan Component. Branch canals, distributaries, minors, field channels etc. would cost about Rs. 100,000 crores. Similarly the cost of power component involving a total of about 30,000 MW part of which would be in Nepal and Bhutan would be of the order of Rs. 80,000 crores. The execution of all these works may take about 40 years including detailed investigation, design etc. Thus the yearly investment on capital works for Himalayan and peninsular component may be around Rs. 8500 crores" (Mohile 1996).

TABLE 4.7

Estimated Benefits from the ILR

Components	Irrigation Benefits (Mha)	Hydropower Benefits (MW)
Peninsular	13	4000
Himalayan	22	30,000
Total	35	34,000

Source: Ministry of Water Resources Studies.

4.8 Review of Proposal of Interlinking of Rivers

4.8.1 Objective

The objective and evaluation criteria do not appear to have been specifically defined. Vague statements often appear. For example, the opening statement in the theme paper on "Interbasin Transfers of Water for National Development—Problems and Prospects" (Mohile 1996) issued by the concerned officials of the Government of India states:

"It is well known that the distribution of India's water resources is highly uneven . . . The basic philosophy of interbasin transfers presumes the need to correct the natural imbalance leading to largely inequitable distribution of water resources . . . Interstate projects and interbasin links would physically make the states interdependent. This will foster day to day cooperation leading to a feeling of oneness. Thus, interbasin transfers would improve the national solidarity."

Interbasin transfer has also been justified based on meeting the increasing demands of water, particularly to obtain food security. Similar vague statements are made by the Task Force on Interlinking of Rivers set up by the Government of India (TFIR 2003).

4.8.2 Analysis of Current Planning Approach

Agriculture demand is dominant, about 70%, and may be examined. The major assumptions in making the estimates of water demand for agriculture by the year 2050 are as follows: (1) the yield of the irrigated food crop by the year 2050 will be 4 tonnes/ha, (2) delta for surface and groundwater will be 0.61 and 0.49 m, respectively, and (3) the cropping intensity will be 160%. The basis of the estimate of yield has not been explained but, perhaps, is on an extrapolative approach. The cropping intensity estimate is also on an extrapolative basis. The estimates of delta are based on marginal improvement on the current figures. In later estimates by the CWC, these estimated improvements in delta were considered to be unattainable, and revised figures of water demand were obtained. The issue of return flows and water quality has been totally ignored.

Suggestions had been made by the author and some U.S. scientists that groundwater supplies can be considerably increased in the Indo-Gangetic Basin by induced groundwater recharge, called the Ganga Water Machine, as discussed in Chapter 5.[3] These have been ignored on the plea that it is not a proven technology. Suggestions have also been made by Chaturvedi (2001) that storage in the peninsular region be worked out by increasing the over-year storage.

4.8.3 Planning Approach

The broad approach adopted in the national perspective is stated to be as follows: (1) Existing uses have been kept undisturbed. (2) Normally water development under the existing legal and constitutional framework is assumed to have taken place fully by the turn of the twentieth century. (3) The development envisaged is within the framework of all the existing agreements between the states. (4) While planning interbasin and interstate transfer of water, reasonable needs of the basin states for the foreseeable future have been kept in view and provided for. (5) Most efficient use of land and water in the existing irrigation and hydropower station has been kept as a principal objective to be attained (Mohile 1996).

Economic, social, and environmental criteria for water resources development and systems planning techniques have been developed (Chaturvedi 1987, 2011a). Proposals, even in the early planning stage, can be profitably analyzed by using these techniques. These, however, do not appear to have been applied for the ILR proposals. As pointed out by several scientists, a simple calculation will show that increased benefits can be better obtained by adopting better irrigation practices such as sprinkler and drip irrigation. It has also been demonstrated that the ILR will have only a negligible beneficial effect on flood mitigation, contrary to claims (Reddy 1991).

4.8.4 NCIWRD Working Group Evaluation of the ILR

A Working Group was instituted by the NCIWRD 1999 itself to study the Interbasin Transfer (ILR) proposal. Its general conclusion and recommendations are as follows:

> It is seen that the country as a whole can produce adequate food for its projected population in 2050. Domestic, industrial and energy water requirements can also be met. This can be done without large scale interbasin transfer of water, provided that necessary intra-basin storages and ground water development are carried out.
>
> There are large disparities in the availability of water between different river basins. The criteria adopted by the Commission for working out surplus or deficits have been given earlier. According to these, cropping and irrigation intensities in all peninsular basins are improved appreciably beyond present situation. All basins are capable of attaining self-sufficiency in food, except for Cauveri and Vaigai by a small margin.
>
> An alternative view can be taken that water as a national resource should be optimally utilized, and where a basin can not make use of all its resources, it should transfer its surplus waters to other basins where it can be used profitably. The technology involved is feasible and within the capability of Indian Engineers. This will undoubtedly lead to more prosperous agriculture in the recipient basins, though at certain economic, social, and environmental costs. Whether these costs are bearable has to be judged by national policy makers. Alternative means of efficient conjunctive water use, and water conservation, in conjunction with conventional intra-basin development have to be evaluated. This can best be done with the help of computer simulation models and systems analysis.

With respect to Himalayan Component, a detailed evaluation could not be made as the reports and studies done so far are classified. Generally speaking the aim is to transfer from rich Brahmaputra and lower Ganga basins towards the west, finally conveying it to water short southern U.P. Haryana, Punjab areas and arid Rajasthan desert. There is also a possibility of transferring water to the peninsular component. The storages and links involved are of very large sizes and lengths: the costs and construction and environmental problems would be enormous. These links should only be taken up if they are considered unavoidable in national interest. Present evidence does not suggest that it is so. For Thar Desert it would perhaps be desirable to promote arid zone low density tree cover as far as possible. The Indira Gandhi Nahar on the west and Narmada canal on the south east, together with practices of desert moisture conservation as developed in Israel can perhaps achieve this limited objective. The need for further expansion of irrigation facilities in this area will have to be examined from all angles including ecological and environmental considerations. (NCIWRD 1998)

4.8.5 Conclusion

As observed by several scientists, the proposal is seriously deficient in many respects (Reddy 1991; Chaturvedi 1998; Singh 2003). As Singh (2003) observed, it can be best described as "a big dream of little logic." In many respects, it can be considered "preposterous" (Chaturvedi 1998).[4] Even from technological considerations, it raises serious criticisms. Constructing canals in the upper reaches of Sivaliks is not acceptable. For instance, the idea of transferring water from Brahmaputra to west and, similarly from the tributaries of Ganga to west, along the foot hills will not be acceptable to any engineer who has done real-life engineering in this area. The classic Cottton–Cautley controversy arose on this issue. The author has designed Ramganga project, which involved construction of only a 5000 ft^3/s canal in the foothills of the Himalayas, and he is aware that it should not be undertaken. As another example, this is why Rao proposed transferring water from Ganga after the confluence of the major northern tributaries and not from Brahmaputra.

4.9 Second India Studies—Water

The Ford Foundation, in 1975, commissioned studies in some areas to "examine the dimension" of the challenge that India will be facing, as a "second India will be added to that which exists" by about 2000. The Second India Studies—Water was undertaken by the author in this context (Chaturvedi 1976). The conclusion of that study is given again as a conclusion of the present study also, as it stands out with even more ferocity. It was concluded that, "The serious predicament that awaits us can be avoided only if there is a revolution in our concepts, organization and capability, and immediate action is taken to meet the critical situation ahead." The review of the state of the water sector in India, as carried out by the NCIWRD 1999, does not display any sign of the urgently needed revolution. Even the NCIWRD 1999 itself did not display how this revolution can be brought about, except for making some pious pronouncements.

4.10 India Water Partnership Study[5]

The Institute for Human Development organized an India Water Partnership (IWP), bringing together a number of interested scientists to produce *India Water Vision 2025* (IWP 2000). The vision represented a "desirable future." Two scenarios were considered for *India Water Vision 2025*: Business as Usual and Sustainable Water World. Numbers of vision elements were suggested. Some of the key vision components include availability of safe drinking water for all near households so that women do not spend much time in fetching water; perception of water used for meeting the basic needs of cooking, drinking, and hygiene as a social good; equity in the use of drinking water; availability of food at affordable prices for the poorest; minimum mortality and morbidity due to water-related diseases; optimum use of water as per agroclimatic conditions; existence of clean rivers, lakes, and other water bodies; minimum flows in rivers and minimal interstate disputes; large dependence on rainwater harvesting; minimum pollution from industries and agriculture; effective regional cooperation in sharing water and energy resources; and effective governance and decentralized management.

The key drivers that influence the outcomes in the above two scenarios were identified and categorized into demographic, social, economic, technological, and international/global. For developing scenarios, the following key drivers were used: population growth; urbanization and emergence of mega cities; economic growth; zero poverty level; and import prices for food grains.

The *vision elements* and *key drivers* were used to develop the two scenarios for 2025. The total estimated demand for water (gross) for 2025 has been identified at 1027 billion cubic meters (BCM). To meet these demands, water availability will have to be increased from about 520 BCM in 1997 to more than 1000 BCM in 2025. This will necessitate an investment outlay estimated at Rs. 5000 billion during the next 25 years or about Rs. 200 billion per year.

Such massive investments in new projects should be planned within the framework of an integrated scheme of river basin development plan. However, before such large projects are planned and taken up for execution, detailed analysis of the various options for meeting sector-wise demands should be made. Such options should include, *inter alia*, the following: (1) recommendations of lifestyles, development paradigms, and attitudes to consumerism; (2) rainfall harvesting for improving soil moisture content; (3) measures for optimum production of crop and its sustainability; (4) watershed development; (5) improving water use efficiency through appropriate technology in irrigation, households, and industry; and (6) recycling and reuse of treated water.

Furthermore, development of water resources projects would require explicit assessment of environment and social impacts. In this context, trade-off between development and environment should be directly addressed, and appropriate decisions should be taken to harmonize the conflicting points of view and development philosophies.

As a review of the study, it may be stated that it misses the modern advances in scientific planning of water resources, as we bring out in the following chapters, and the observations remain pious statements.

4.11 World Bank Study[6]

The World Bank undertook a study of the strategic challenges facing the water sector in India (Briscoe and Malik 2006). It was based on several commissioned papers. Some observations and recommendations are as follows.

India faces a turbulent future. The current water development and management system is not sustainable; unless dramatic changes are made—and made soon—in the way in which government manages water, India will have neither the cash to maintain and build new infrastructure nor the water required for the economy and the people.

Large investments have been made in India's water structure, but India still needs a lot more water infrastructure. Whereas arid-rich countries (such as the United States and Australia) have built over 5000 m^3 of water storage per capita, and middle-income countries such as South Africa, Mexico, Morocco, and China can store about 1000 m^3 per capita, India's dams can store only about 30 days of rainfall, compared with 900 days in major river basins in arid countries. A compounding factor is that there is every indication that the need for storage will grow because of the climate change.

Similarly, there is a considerable shortage on hydropower structure. Whereas the industrialized countries harness over 80% of their economically viable hydropower, the figure is only about 20% for India.

Besides the problem of providing adequate quantities of water, growing populations, cities, and industries are putting great stress on the environment. Many rivers—even very large ones—have turned into fetid sewers. India's cities and industries need to use water more effectively.

Global experience shows that the returns to investments in water infrastructure follow a typical pattern. During the first development stage, the challenges are predominantly engineering in nature. As the infrastructure platform is built, challenges of maintenance, operation, and management start to emerge. The uni-functional ("build") and uni-disciplinary ("engineering") bureaucracy in India adopted the command-and-control philosophy of the early decades of independence, seeing users as subjects rather than partners or clients. The Indian state water apparatus still shows little interest in the key issues of management stage—participation, incentives, water entitlements, transparency, entry of private sector, competition, accountability, financing, and environmental quality.

Evidence abounds of the inadequacy of the state water machinery to address even the problems of provision of public irrigation and water supply services. User charges are negligible, resulting in lack of accountability and insufficient generation of revenue even for operation and maintenance. The gap between tariff and value of irrigation and water supply services has fueled endemic corruption. Staffing levels are 10 times international norms, and most public funds are now spent feeding the administrative machinery and not maintaining the stock of infrastructure or providing services. There is an enormous backlog of deferred maintenance.

The sector is facing major financial gaps. This can only be met by a combination of methods that include greater allocations of budgetary resources, more efficient use of those resources, and greater contribution from water users.

These difficulties have been tried to be met by people by muddling through. However, this is not adequate and safe. It will not be able to meet the huge future challenges. At the same time, Indian society is changing in many profound ways. Industries and cities (which both require and produce wastes) are growing rapidly. Rural life is changing, with more than half of the people in Punjab and Haryana no longer engaged in agriculture. In addition, agriculture itself is evolving. In a growing number of areas, high-value crops are displacing low-value food grains, farmers are investing heavily in drip irrigation, and so on. As incomes rise—100,000 people are joining the middle class every day—people are becoming more concerned with environmental quality. The net effect is that demands for and on water resources are changing substantially, with the effects especially acute in the high-growth regions, most of which are water scarce.

Confronted with this reality of limited supplies, and growing and changing demands, the need is obviously for a management framework that stimulates efficiency and that facilitates voluntary transfer of water as societal needs change. The traditional command-and-control and construction instruments of the Union and State water bureaucracies address neither of these imperatives. The economic and social costs of these rigidities are large. A central element of a new approach must be that the users have well-defined entitlements to water. The broader messages are that the economic ideas of

1991 reforms must be drilled down from the regulatory and financial sectors into real sectors (including the water sector) if India has to have sustainable growth, and the role of the Indian water state from that of a builder and controller to a creator of an enabling environment and facilitator of the actions of water users, large and small.

An important manifestation of the breakdown of the current system is the growing incidence and severity of water conflicts—between states, between cities and farmers, between industry and villagers, between farmers and the environment, and within irrigated areas. The results have been serious economic and fiscal damages.

India needs a reinvigorated set of public water institutions, which are built on the following imperatives:

1. Focusing on developing a set of instruments (including water entitlements, contracts between providers and users, and pricing) and incentives that govern the use of water
2. Stimulating competition in and for the market for irrigation, water, and sanitation services
3. Empowering users by giving them clear, enforceable water entitlements
4. Ending the culture of secrecy and making transparency the rule
5. Introducing incentive-based, participatory regulation of services and water resources
6. Putting the sector on a sound financial footing
7. Investing heavily in the development of a new generation of multidisciplinary water resource professionals
8. Making the environment a high priority
9. Making local people the first beneficiaries of major water projects

India is rapidly approaching the end of an era in which society could "get by" despite the fact that the government (1) has performed poorly where it has engaged (in service delivery) and (2) has abandoned major areas where government engagement is critical (such as groundwater management, conflict resolution, establishing and managing water entitlements, and the financing of public goods such as flood control and wastewater treatment).

There are two main corollaries to this diagnosis. First, a major push is needed—by government and by users working together—to bring abstractions in line with recharge. Whereas traditional technologies such as rainwater harvesting and tanks can play an important role, they also create new and additional demands that often clash with existing uses, and they sustain the wishful thinking that supply-side options (both large and small scale) are what will "solve the problem." The simple fact is that in many parts of India, demand will have to be brought down to match the sustainable supply. Global experience shows that this difficult and essential task will require a partnership between users and government to form empowered aquifer user associations, to formalize water entitlements that are consistent with sustainable yield of the aquifer, and to develop transparent information and decision support systems. So far, the approach of the water apparatus has been to promulgate laws and policies, most of which are not implemented. Here, an approach that begins with acknowledgment and respect for the private interests of the individual farmers will be far more successful than approaches that resort to command and control, or which are based on a communitarian ideal. The longer this adjustment takes place, the more costly and difficult it becomes.

Second, the end of the era of massive expansion of groundwater use is going to demand greater reliance on surface water systems. This is going to require recuperation of the large stock of dilapidated infrastructure and large-scale investment in public structures of all scales (for provision and distribution of surface water supplies, but also treatment of wastewater). It is going to require a dramatic transformation in the way in which public water services are provided to farmers, households, and industries, in which the watchword are entitlements, financial sustainability, accountability, competition, regulation, and entry of alternatives to government provision, including cooperatives and the private sector.

India faces the challenge with many assets and some liabilities. The assets include citizens, communities, and a private sector that have shown immense ingenuity and creativity—attributes that are critical for the new era of water management. The major liability is a public sector that rests on the laurels of an admirable past, but is not equipped with the central tasks that only the government can—developing an enabling legal and regulatory framework; putting into place entitlement and pricing practices that will provide incentives for efficient, sustainable, and flexible use of water; forming partnerships with communities for participatory management of rivers and aquifers; providing transparent information for use in managing and monitoring the resource and services; stimulating competition among providers through benchmarking and the entry of private sector and cooperative providers; regulating both the resource and services; and financing true public goods, such as flood control and waste water treatment. Figure 4.2 provides a schematic sense of the necessary "next stage" in the evolution of water management in India.

In the eyes of many, the idea of such modern accountable "India water system" is a fantasy given the dismal performance of the Indian state on water matters in recent decades and the broader challenge of governance. Others point to "the hollowing out of Indian state . . . the growing middle class exit from public services . . . and the inability to grapple with many long term challenges facing the country." The glass is, of course, always half empty. However, it is half full, too. There are some important signs that change is needed, there are political leaders who are starting to grapple with these realities, and there are a few states that are taking the important first steps down this long and winding road.

India is fortunate, too, in that it is not the first country in the world to face these (daunting) challenges. The experiences of other countries suggest that there are a set of "rules for reformers" in undertaking such a transition. These rules include the following:

1. Initiate reform when there is a powerful need and demonstrated demand for change.
2. Involve those affected and address their concerns with effective, understandable information.
3. If everything is a priority, nothing is a priority—develop a prioritized list of reforms.
4. Pick the low-hanging fruit first—nothing succeeds like success.
5. Keep your eyes on the ball—do not let the best become the enemy of the good.
6. Be aware that there are no silver bullets.
7. Do not throw the baby out with the bathwater.
8. Treat reform as a dialectic, not mechanical process.
9. Understand that all water is local and each place is different—one size will not fit all.
10. Be patient, persistent, and pragmatic.

Ensure that the reforms provide returns to politicians who are willing to make changes.

4.12 Review of the World Bank and Other So-Called Expert Studies[7]

The World Bank study was presented a bit in detail to bring out as to what these studies by so-called foreign experts mean. It will be seen that besides some general observations and managerial suggestions, no scientific or technological advancements are proposed. Reference to another important study (Wilderer 2011) may also be made. It also suffers from the same limitation, except Lenton's elucidation of IWRM, which will be brought out a bit in detail, in terms of considering the much emphasized IWRM. It will be further shown in Chapter 6 that we have to extend the concepts further in terms of Sustainable Environmental Systems Management, as proposed by the author.

The central issue is that, generally, these people have no experience of real-life engineering and also of Indian conditions. This will come out sharply as we bring out, in the following pages, the possibility of revolutionizing the development and management of India's waters. This observation is made to emphasize a central fact that we have to undertake our development ourselves, leading to the highest levels.

4.13 Reflections on Governance

Iyer (2003, 2007) has made studies regarding water governance, politics, and policy, an area generally neglected. The subject will be considered in Chapter 7.

4.14 Integrated Water Resources Management

The concept of IWRM was propounded in the Dublin Conference in the 1990s. Much discussion has taken place. Recently, the Technical Committee of the Global Water Partnership (GWP) defined it as "a process which promotes the coordinated development and management of water, land and related resources, in order to maximize the resultant economic and social welfare in an equitable manner without compromising the sustainability of vital ecosystems."

As Lenton (2011) states, following GWP Technical Committee (2009) . . . IWRM is the way in which . . . water can be managed to achieve the objectives of sustainable development, and an approach that reflects the need to achieve a balance among economic efficiency, social equity, and environment sustainability. Analyzing integrated approaches to water resources management at different levels (from small watersheds to basins, agricultural systems, and national and global policy making), which have been practiced for some time and about which there is a considerable body of knowledge. Building on these analyses, IWRM is not to be seen as a single approach but as a whole range of approaches to manage water and related resources—a meta-approach or meta-concept, as it were, which both transcends the various levels of decision making at each level. The chapter then provides some historical perspective on the evolution of IWRM, concluding with a summary of recent assessments and critiques of concepts.

Although the concept of IWRM is applicable to a variety of contexts, this chapter focuses on management of water in the context of development, that is, on the management of water resources to advance sustainable development and reduce poverty.

Two perspectives that characterize all writings on the subject are (1) concerned only with water, even when diverse socioeconomic–environmental aspects are considered and (2) "on the management of water resources to advance development and reduce poverty."

Our point of departure, as brought out in Societal Environmental Systems Management, is from both these two basic concepts (Chaturvedi 2011a). First, we consider not only the management of water with associated environmental components but also the management and modernization of the associated societal system. This leads to a tremendous contribution to socioeconomic development and sustainability. Second, we do not limit ourselves to eradication of poverty but conjunctively consider raising the society to the highest socioeconomic levels. It emphasizes social justice in fundamental basic terms and does not constrain us to the historic concept of a poor backward Third World as a continuing human domain, which is a deeply embedded concept in the First World thinking, as was the classical Brahmanical concept. The development up to the much higher socioeconomic levels puts enormous pressure on the environment. We show in Chapter 5, and even developed some novel technologies, that it is eminently feasible.

We make another advancement. We emphasize that the practice of IWRM or more appropriately, Societal Environmental Systems Management, has to be undertaken through participatory management. For this, and for developing the proper policy, participatory system dynamic modeling will have to be undertaken (Chaturvedi 2011a).

4.15 Recent Modification by GOI of Water Demand Estimates

The XIth Plan (2008–2012) brings out the latest official thinking in the subject. Some observations are reproduced.

INTRODUCTION

2.1 Sustainable development and efficient management of water is an increasingly complex challenge in India. Increasing population, growing urbanization, and rapid industrialization combined with the need for raising agricultural production generates competing claims for water. There is a growing perception of a sense of an impending water crisis in the country.

Some manifestations of this crisis are:

- There is hardly any city which receives a 24-hour supply of drinking water.
- Many rural habitations which had been covered under the drinking water programme are now being reported as

having slipped back with target dates for completion continuously pushed back. There are pockets where arsenic, nitrate,
and fluoride in drinking water are posing a serious health
hazard.

- In many parts, the groundwater table declines due to overexploitation imposing an increasing financial burden on farmers who need to deepen their wells and replace their pump sets
and on State Governments whose subsidy burden for electricity supplies rises.
- Many major and medium irrigation (MMI) projects seem to
remain under execution forever as they slip from one plan to
the other with enormous cost and time overruns.
- Owing to lack of maintenance, the capacity of the older systems seems to be going down.
- The gross irrigated area does not seem to be rising in a manner
that it should be, given the investment in irrigation. The difference between potential created and area actually irrigated
remains large. Unless we bridge the gap, significant increase in
agricultural production will be difficult to realize.
- Floods are a recurring problem in many parts of the country. Degradation of catchment areas and loss of flood plains
to urban development and agriculture have accentuated the
intensity of floods.
- Water quality in our rivers and lakes is far from satisfactory.
Water in most parts of rivers is not fit for bathing, let alone
drinking. Untreated or partially treated sewage from towns
and cities is being dumped into the rivers.
- Untreated or inadequately treated industrial effluents pollute
water bodies and also contaminate groundwater.
- At the same time water conflicts are increasing. Apart from
the traditional conflicts about water rights between upper
and lower riparians in a river, conflicts about quality of water,
people's right for rainwater harvesting in a watershed against
downstream users, industrial use of groundwater, and its
impact on water tables and between urban and rural users
have emerged.

2.2 India with 2.4% of the world's total area has 16% of the world's population, but has only 4% of the total available freshwater. This clearly
indicates the need for water resource development, conservation, and
optimum use. Fortunately, at a macro level India is not short of water.
The problems that seem to loom large over the sector are manageable
and the challenges facing it are not insurmountable.

AVAILABILITY OF WATER RESOURCES

2.3 The water resource potential of the country has been assessed from time to time by different agencies. The different estimates are shown in Table 2.1. It may be seen that since 1954, the estimates have stabilized and are within the proximity of the currently accepted estimate of 1869 billion cubic metre (bcm) which includes replenishable groundwater which gets charged on an annual basis.

TABLE 2.1

Estimates of Water Resources in India

Agency	Estimate	Deviation from 1869 bcm
First Irrigation Commission (1902–03)	1443	–23%
Dr. A.N. Khosla	1673	–10%
Central Water and Power Commission	1881	+0.6%
National Commission on Agriculture	1850	–1%
Central Water Commission (1988)	1880	+0.6%

UTILIZABLE WATER RESOURCES POTENTIAL

2.4 Within the limitations of physiographic conditions, socio-political environment, legal and constitutional constraints, and the technology available at hand, the utilizable water resources of the country have been assessed at 1123 bcm, of which 690 bcm is from surface water and 433 bcm from groundwater sources (CWC, 1993). Harnessing of 690 bcm of utilizable surface water is possible only if matching storages are built. Trans-basin transfer of water, if taken up to the full extent as proposed under the National Perspective Plan, would further increase the utilizable quantity by approximately 220 bcm. The irrigation potential of the country has been estimated to be 139.9 MH without interbasin sharing of water and 175 MH with interbasin sharing.

2.5 While the total water resource availability in the country remains constant, the per capita availability of water has been steadily declining since 1951 due to population growth. The twin indicators of water scarcity are per capita availability and storage. A per capita availability of less than 1700 cubic metres (m^3) is termed as a water-stressed condition while if per capita availability falls below 1000 m^3, it is termed as a water scarcity condition. While on an average we may be nearing the water-stressed condition, on an individual river basin-wise situation, nine out of our 20 river basins with 200 million population are already facing a water-scarcity condition. Even after constructing 4525 large and small dams, the per capita storage in the country is 213 m^3 as against 6103 m^3 in Russia, 4733 m^3 in Australia, 1964 m^3 in the United States (US), and 1111 m^3 in China. It may touch 400 m^3 in India only after the completion of all the ongoing and proposed dams.

ULTIMATE IRRIGATION POTENTIAL (UIP)

2.6 The demand for irrigation water in India is very large. However, the limits to storage and transfer of water restrict the potential for irrigation. UIP reassessed by the Committee constituted by the MoWR in May 1997, the potential created, and the potential utilized up to the end of the Tenth Plan.

2.7 The assessment of UIP needs to be periodically reviewed to account for revision in scope, technological advancement, inter-basin transfer of water, induced recharging of groundwater, etc. The creation of irrigation potential depends upon the efficiency of the system for delivering the water and its optimal use at the application level. With the modern techniques of integrating micro irrigation with canal irrigation, as has been done in the case of the Narmada Canal Project, Rajasthan, the UIP can further be increased. Similarly in the case of groundwater, innovative methods of recharging the groundwater and also storing water in flood plains along the river banks may enhance the UIP from groundwater to more than 64 MH.

WATER FOR NATURE

2.8 The question of a trade-off between competing claims on water becomes most important in the context of ecological requirement. The National Water Policy (NWP) places ecology in the fourth place in the order of priorities for water use. Yet, there is a general agreement amongst all that any water diversion Water Management and Irrigation 45 needs to take care of river ecosystem downstream. The problem is of quantifying the Environment Flow Releases (EFR) that is the flow required for maintaining ecosystems. Usable water will be reduced to that extent. During 2004–05, the Ministry of Environment and Forests (MoEF) appointed a committee headed by Member, Central Water Commission (CWC), to develop guidelines for determining the EFR. The committee submitted its report in 2005. Depending on what the final accepted recommendation is, the minimum flow required for maintaining the river regime and environment will be decided and considered in water resources development and management.

CLIMATE CHANGE AND UNCERTAINTY
IN WATER AVAILABILITY

2.9 The threat of climate change is now considered an established fact. General Circulation Models simulate the behavior of the atmosphere and paint 'what if' scenarios for various levels of greenhouse gas emissions. Using these models the weather experts have predicted that global warming will intensify the hydrologic cycle; more intense rainfall will occur in fewer spells; floods and droughts both will become more intense; the floods will be more frequent; the rainfall will shift toward winter; and there may be a significant reduction in the mass of glaciers, resulting in increased flows in the initial few decades but substantially reduced flows thereafter.

2.10 The MoWR has already initiated some studies in co-operation with research institutions and reputed academic institutions to assess the impact of climate change on water resources.

2.11 The hydrologists are yet to translate what climate change means for the water availability, its distribution in time and space, and changes in demand. An increase in mean temperatures would increase the energy flux for evapo-transpiration. The increased potential evapo-transpiration in the forests could trigger major changes in the environment, and it would result in an increased crop water requirement in the farms. The changes in seasonal temperatures could change the crop seasons. Enough data is now available to paint 'what if' scenarios for different possibilities, and to formulate some tentative plans to respond to these possibilities.

2.12 In the post-climate change scenario, systems that are more resilient will fare better than systems that are less resilient. Engineering infrastructure that enables the water managers to store and transfer water with greater certainty can reduce the impact of uncertainty. Climate change considerations need to be factored in as we plan water resource infrastructure.

WATER REQUIREMENT

2.13 The requirement of water for various sectors has been assessed by the National Commission on Integrated Water Resources Development (NCIWRD) in the year 2000. This requirement is based on the assumption that the irrigation efficiency will increase to 60% from the present level of 35–40%. The Standing Committee of MoWR also assesses it periodically. These are shown in Table 2.2.

TABLE 2.2

Water Requirement for Various Sectors

	Demand in km³ (or bcm)					
	Standing Subcommittee of MoWR			NCIWRD		
Sector Water	2010	2025	2050	2010	2025	2050
Irrigation	688	910	1072	557	611	807
Drinking water	56	73	102	43	62	111
Industry	12	23	63	37	67	81
Energy	5	15	130	19	33	70
Others	52	72	80	54	70	111
Total	813	1093	1447	710	843	1180

WATER RESOURCES DEVELOPMENT AND USE: IRRIGATION

HISTORICAL DEVELOPMENT

2.14 The planned development of irrigation sector started in a big way since the First Five-Year Plan (1951–56). New projects were taken up in the Second Five Year Plan, the Third Five Year Plan, and the Annual Plans 1966–69. During the Fourth Five Year Plan emphasis was shifted to the completion of ongoing schemes. The widening gap between potential creation and utilization was felt in the Fifth Plan (1974–78) and accordingly Command Area Development (CAD) programme was launched. The Annual Plans 1978–80 and the Sixth Plan witnessed new starts and then the focus was shifted toward completion of irrigation projects. By the end of the Eighth Plan (1996–97), central assistance was provided under AIBP to help the State Governments in early completion of the projects.

2.15 Although plan expenditure on irrigation has increased from Rs. 441.8 crore in the First Plan to Rs. 95743.42 crore (outlay) in the Tenth Plan, the share in total plan expenditure has decreased from 23% in the First Plan to 6.3% in the Tenth Plan. The trends in change of percent of total plan expenditure on irrigation sector are shown in Figure 2.1.

2.16 The anticipated irrigation potential created up to March 2007 is 102.77 MH, which is 73.46% of the UIP of 140 MH. MMI projects have an UIP of 58.47 MH against which irrigation potential created is 42.35 MH. MI potential created is 60.42 MH against the UIP of 81.43 MH. The irrigation potential creation and its corresponding utilization during the plan periods is given in Annexure 2.1.

2.17 The gross irrigated area in the country is only 87.23 MH. With an average irrigation intensity of 140%, the actual net irrigated area is likely to be around 62.31 MH, which is only 43% of the net sown area of the country (142 MH). Even after achieving the UIP of 139.89 MH and considering the average irrigation intensity of 140%, the ultimate irrigated area in the country would be only 70% of the net sown area . . .

4.16 Overview of Current Official and Other Studies

Studies of the current official and others bring out one central fact. They are confined to the historic concept of the development of India's waters. Any suggestions are marginal comments. We will show that a revolution in concepts, technology, and management is possible and is urgently needed.

4.17 Conclusion

We believe that India deserves some revolutionary advances in water science and engineering. An associated question arises as to why this has been lacking. Both will be addressed in the following pages.

Notes

1. NCIWRD 1999.
2. The approach and perception have been stated in Commission's words but cannot be put under quotes because an attempt has been made to summarize them.
3. The challenge of storing the monsoon runoff was raised in an Indo-U.S. Science Academy Seminar in New Delhi in the early 1970s attended by Chaturvedi and Revelle. Several suggestions were made. The author and Revelle pursued the idea and studied to different levels of detail. The scheme was discussed informally by the author with Dr. K.L. Rao, Minister of Irrigation, GOI, with whom he had the privilege of working closely. He approved it and appointed a committee to carry out preliminary tests. This was done and the

feasibilty was established. Detailed studies were proposed to be undertaken but he later left.

Revelle and Chaturvedi independently suggested lowering of the groundwater all over the plains, though the approach was slightly different (Revelle and Herman 1972; Lakshminarayana and Revelle 1975; Revelle and Lakshminarayana 1975; Chaturvedi et al. 1975; Chaturvedi and Srivastava 1979; Srivastava 1976). Revelle gave the name Ganga Water Machine and accepted Chaturvedi's approach, and this is why it is called Chaturvedi Ganga Water Machine. The two worked closely for the implementation of the proposal. A workshop of Indian and U.S. scientists was held in 1981, and the subject was discussed in detail. Field visits of the workshop scientists were organized. The scheme was approved by the workshop scientists. It was decided that further studies and field tests should be carried out. According to the terms of international scientific collaboration, this required concurrence of the Government. The subject was, therefore, discussed in an informal meeeting of concerned Indian and U.S. scientists (Roger Revelle, Peter Rogers, and the author) with the then Member, Planning Commission (Dr. M.S. Swaminathan), attended by the then Secretary, Government of India, Water Resources (Shri C.C. Patel). Shri Patel promised to get the field tests carried out. Unfortunately, these were not carried out. The NCIWRD notes that they should be carried out urgently.

4. The author was invited by the Parliamentary Standing Committee on Agriculture to give his views on the subject of utilization of national waters for irrigation with special reference to the subject of interlinking of rivers, inviting a comprehensive memorandum on the subject. The Committee decided to hear the views in person, which were presented. The Commission wanted to hear the views further and another personal appearance was made.

5. IWP 2000.

6. Briscoe and Malik 2006.

7. John Briscoe and several of the authors of the Treatise of Water Science are close friends of the author. However, it becomes necessary to make these scientific observations because one of the important issues that we are trying to emphasize is necessarily of indigenous studies, as our conditions cannot be understood by foreign academicians unless they are involved in real-life planning in India.

5

---

*Revolutionizing Development of
India's Waters—New Concepts,
Policies, and Technologies*

5.1 Introduction

In our judgment, a fundamental but logical revolution is required in our thinking about India's development. India, representing the largest group of humanity (about 17%), has to achieve, logically, her due place in human community. This is the first step in our thinking about water. To establish this basic thought, not just as a dream but also as a reality, we will demonstrate that several novel technologies are possible, which can revolutionize the water scene of India and concomitantly the development of India. Climate change poses additional serious challenges, which have to be dealt with integrally with developmental planning. We will provide some ways to meet this challenge also.

We briefly review the current official perspective at the outset to bring out that a revolution is needed in the concepts, policy, technology, and planning process, and then we unfold them. These ideas have to be tested and advanced through an appropriate level of planning.

5.2 Study of Current Policy and Management

The development and management of India's waters and proposed advances by the NCIWRD (1999) have been discussed in Chapters 3 and 4, respectively. Certain concepts have dominated thinking of the water resources development of India. We will demonstrate that these concepts are not valid. The current thinking may, therefore, be first overviewed briefly.

Management of water is crucial for agriculture in India on account of the climatic–hydrologic characteristics. The concern of the people was, merely, to get by. There are two agricultural seasons: monsoon period (Kharif) and post-monsoon winter period (Rabi). The rainfall is concentrated in a few days, in heavy precipitation during the monsoons months. In the monsoon agricultural season (Kharif), the farmer depended on rains. They were often deficient or untimely, and famines were frequent and intensive (Bhatia 1967). However, that was considered God's wish, judging his deeds. All he could do was to pray. During the winter season (Rabi), rains are a token, and he would try to supply water to crops in whatever way locally possible. Along rivers, water was tried to be diverted in the earlier sowing season when there were enough low flows to enable their diversion for a while. Later, he had to depend on rains if and when a sprinkling became available.

It is important to note that the environmental characteristics of water were entirely different in Western Europe and the Eastern United States, the developing Western world. As Landes (1999, p. 18) points out, "They were assisted here by a relatively even rainfall pattern, distributed around the year and rarely torrential: 'it droppeth as the gentle rain from heaven.' This is a pattern found only exceptionally around the globe. . . . This dependable and equable supply of water made for a different pattern of social and political organization from the prevailing riverine civilizations. Along rivers, control of food fell inevitably to those who held the streams and the canals it fed. Centralized government appeared early, because the master of food was the master of people. The biblical account of Joseph and Pharaoh tells this process in allergy. In order to get food, the starving Egyptians gave up to Pharaoh first their money, then their livestock, then their land, then their persons (Genesis 47: 13–22). Nothing like this was possible in Europe." As he continues, "By the thirteenth century, China thus had the most sophisticated agriculture in the world, India being the only conceivable rival" (Landes 1999, p. 26).

A constellation of technological, policy, financial, political, and practical factors dictated the policy of development of the water resources in the British period in the context of the Indian environmental characteristics. Because they had no experience of developing irrigation, in the early period, they just repaired the existing works. Attempts at development of canal irrigation took place gradually. Cautley (1860), who constructed the first major irrigation work, the Ganga Canal, went to Italy to study how canals were constructed on the tiny river Po.

The objective remained the same—stabilization of the sustenance agriculture and minimization of famines. As shown by the characteristics of the river runoffs (see Figure 5.1), it was not considered possible to manage the monsoon period flows. An attempt was made to improve the Rabi period water supply to crops through better diversion and canal arrangements. The total water availability during the period was considerably short of the total demand. The diverted waters were, therefore, tried to be distributed over as large an area as practically and economically possible, trying to distribute the limited water availability to a reasonably large area so that benefits were

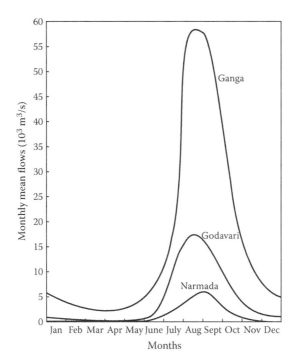

FIGURE 5.1
Typical hydrograph of some major rivers of India.

available to as large a number of people as possible, and thereby minimizing hardships from possible famines. Stabilization of the primitive agriculture, rather than productive agriculture, and protection from famines, as in the past, were the determinants of the activities. The agriculture and irrigation policy and technology remained the same as historically practiced, except that bigger and technologically more sophisticated structures were gradually constructed. There was no concept of socioeconomic development, and poverty continued to be a problem for a lot of people.

Three waterings were attempted to be provided, but as the low flows dwindled, the area served by the second and third waterings would be reduced. No storage works were developed. The total irrigated area in India was increased from the historic 6.9 to 22.6 Mha by 1951, the year the British left. However, the increasing population and exploitatory policy of the foreign government created increasing strain on the economy, and per capita food availability continued to drop (Figure 3.1).

Highest emphasis has been given to water resources development after Independence. The recent 1943 famine of Bengal was still very fresh. India has achieved one of the highest areas of irrigated agriculture in the world, an impressive figure of about 100 million acres, by 2009. Large canals and impressive dams have been constructed. There has been an impressive and increasing groundwater development recently.

There is, however, another side of the picture. As one reviews the current development of water resources of India, including the proposals made for future development (NCIWRD 1999, XIth Plan), or even observations made by independent observers as brought out in Chapter 3, a significant important observation emerges. The concepts, objective, policy, and technologies that are being adopted currently or are proposed to be adopted in future developments are almost the same as those developed since prehistoric times, particularly the British, except that the scale has been enlarged several times. The objective still is meeting the food demands of the country. The perspective still is the world's poorest agriculture, with estimates of yield in 2050 that have already been surpassed in China and much less in emerging advanced developing countries like Korea. The central feature continues to be using the low flows after the monsoons are over. The focus has always been and continues to be on the low flows, and the entire development is woven around it. Since Independence, attempts to store the monsoon waters through construction of dams have been made only recently, and still continue, but the development policy remains to use these additional waters to extend irrigation on the pattern already established historically based on the focus on low flows to stabilize the sustenance agriculture. Even the groundwater developed in the public sector was proposed to be used to extend irrigation on the same principle, providing an apology of irrigation. The focus has been on extending irrigation without bothering about the quality of irrigation or the environmental impacts or economic development. This is reflected in the fact that though India has achieved one of the highest areas of irrigated agriculture in the world, an impressive figure of about 100 million acres, the yields are one of the lowest even in developing countries (WR 1996–97). As already observed, the yields in China are almost twice those in India, and the yields in the industrialized countries are almost three times those in India. Moreover, as stated earlier, drinking water and sanitation position has been neglected and is shameful. With the diversion of the low flows and increasing urbanization and industrialization, which are still in the early stages, the rivers have already been turned into open sewers. A very depressing future can be visualized as the heavy demands of increasing population and economic development press on the environment if the current policies are continued to be followed.

The perspective of environment is another matter of concern. In contrast to the earlier neglect of the environmental perspective, mention of the environment has been made in NCIWRD (1999) and latest Plan documents, but it is a passing reference. The subject is now being considered at a much more advanced level in terms of sustainable development (BSD 1999; Chaturvedi 2011a). The perspective is particularly important for India because a very large socioeconomic development will be taking place at a very rapid rate.

The central fact that emerges is that the thinking has not broken much with the past colonial perspective and has not addressed creatively the emerging challenges.

5.3 Perspective of Development

Even more importantly, the perspective development has not broken with the colonial concept of India's development, considered to mitigate poverty, even though the objective of the Independent India development is entirely different. As proclaimed in the first breath of Independence by Jawaharlal Nehru in his famous Tryst with Destiny speech on August 15, 1947, a commitment was made to bring the vast human community to achieve the rightful place in the human community. This has been emphasized in the Plan Document and the speeches of Honorable Prime Minister, Manmohan Singh. Eradication of poverty is a priority, as expressed in terms of the policy of "inclusive development," but the objective is to join the ranks of humanity at the highest level and at the earliest.

However, this spirit is not reflected in the management of water, which is crucial for development. In our judgment, a fundamental correction is, therefore, needed in the perceptions and is the cornerstone of our suggestions. Starting from the current lowly place, it means tremendous commitment and a revolution in our efforts, as one has to be a step better, if equality has to be achieved, when one is far behind. The central objective is attainment of due position as one of the largest components of humanity and, as a corollary, leading position, or in other words, global leadership. This has to be attempted to be achieved in all fields, particularly regarding environmental management, of which water is a dominant component. Another corollary is the rapid eradication of poverty at the earliest. This is not a chauvinistic proposition but a logical corollary from the basic premises of justice to humanity, with India representing the largest component of humanity, about 17%, as officially propounded.

5.4 Challenges

In formulating the water policy, we take an entirely different approach than that adopted by the governmental agencies and the NCIWRD (1999) entrusted with the task. They essentially extrapolated the future from the past. We try to look at the future as it is likely to unfold under the pressure of population explosion and the proposed economic development of the country. The current trends confirm this perspective.

An important issue to be borne in mind is that the trajectory of development and its implications on environment for the developing countries, of which India with about 17% of the global population is a dominant component, along with China, will be entirely different than those of the currently industrialized countries as their development took place. They faced the challenge very gradually at their terms of development. Their population was much smaller. The entire world was at their disposal. Science and technology had not developed that much as India and China find them. As the scientific term is, the initial and boundary conditions for India, China, and the Western countries, regarding socioeconomic–environmental development, are vastly different. Furthermore, the rate of change for these late comers will be much higher than that achieved by the currently developed countries in view of the societal–technological advances. The challenges, social and environmental, are also much higher.

Environmental management is a crucial part of sustainable development. A serious challenge is posed. India has, in terms of per capita, a very small environmental availability, only about 4% against a population of about 17%. Therefore, the highest level of science, technology, and, above all, commitment is required in the management of the environmental system, a part of which is water.

Focusing on water, two characteristics stand out. First, the per capita availability is, comparatively, very limited. For instance, the figures of per capita availability of freshwater resources (cubic meters per capita) for India and the United States are 2167 and 9259, respectively. This is just a vindication of the characteristic of the comparative overall scarcity of environment. In addition, in view of the climatic–hydrologic characteristics, development of water for agriculture is a central challenge, requiring an overwhelming component of the development of water for meeting the agriculture requirements in India. Currently, it accounts for about 80% of the total water use. Not much effort is required, in large parts of the industrialized world of Europe and the United States, regarding management of water for agriculture, and their requirement for the agricultural sector is only about 10% of the total development.

5.5 Future Perspective

The proposed approach to meet the formidable challenge of managing the environment is, therefore, not only in terms of meeting the emerging challenges but also to manage the challenges themselves, trying to foresee them. We, therefore, first focus on future societal challenges and their perspective of sustainable management. Next, conjunctively, we also propose sustainable management of the environment, as well as water, which is a part of it, in this context.

5.5.1 Habitat, Spatial, and Infrastructural Development

People organized themselves over time to manage their natural resource needs. In view of the differing environmental characteristics, the pattern of socioeconomic development was different in the currently industrialized countries and India. In the Western world, the organization was woven around a manor. In India, it was woven around a village, wherein someone would come over and collect land revenue.

Urbanization took place as population increased and economic activities developed in the Western countries. They added land from all over the world for populations to migrate. The scene in India is entirely different. As population increases and increasing economic activities are undertaken, an enormously heavy pressure will be brought to bear on the society and the environment at a very fast rate. It will not be possible for the increasing population to meet their needs at the rural level, and migration to urban centers, where opportunities appear to exist, will naturally take place. The monumental explosion that is taking place in the metropolitan centers, Bombay, Delhi, Calcutta, Chennai, and gradually at the state metropolitan centers, can already be witnessed. For reasons briefly indicated earlier, they are of an entirely different order of magnitude than those faced by other currently industrialized countries as their development took place. The challenge of urbanization in India and China is one of the most difficult challenges ever faced in this sector (Bairoch 1988). Provision of the infrastructure of housing, transportation, and water for drinking and sanitation are very difficult issues.

Habitat is the first issue in human activities and we start with it. Thus, the urgent need is to deal with the issue of enabling and encouraging people to undertake socioeconomic development in the rural areas itself. It has to be noted that the central issue of rural development is not merely producing sufficient food and mitigating poverty, but bringing the people in the mainstream of human development at international levels. It is, therefore, necessary for the developing countries, as they embark on their trajectory of development, that balanced spatial development is emphasized, first and foremost, for the well being of people and the environment. The advances in modern technology of transportation, communication, and energy offer opportunities that were not available to the currently industrialized Western world when their development took place. The simple basic principle that emerges is that rural development has to be undertaken so as to encourage balanced economic and spatial development. No doubt, it is easier said than done, but the Government of India has accepted the principle and is giving the highest attention to the subject in terms of comprehensive development through a policy and plan called Bharat Nirman Yojana (literally, India Construction Plan). A PURA (providing urban amenities to rural areas; PURA means full in Hindi) model has been proposed by the former President, Kalam, who is an eminent scientist (Gandhi 2005). According to this concept, all types of connectivity and facilities—transportation, information technology, energy, housing, education, and health—are proposed to be provided so as to bring the rural people in the main stream of development. This will also lead to stopping and even reversing rural–urban migration. The activities have to be developed, but it is interesting to note that major industrial houses have started taking advantage of the proposed policies. However, the issue is not merely of rural markets, as the current concept is, but of the development of the rural world. As recent reports suggest, the idea is vigorously being practiced in China.

We have further extended the PURA model to PURAC model. The issue is not merely providing the urban amenities but developing the urban culture, which has been the key to development (Bairoch 1988; Landes 1999; Chaturvedi and Chaturvedi, in preparation). Thus, rural development has to be undertaken in a regional perspective to develop appropriate capabilities and to usher the urban culture for socioeconomic activities.

It is the only way to provide even the basic amenities of drinking water and sanitation to the rural and urban areas, which is the first priority in water resources development. What it means, in terms of the policy of water resources development, is that the river flows have to be conserved, first and foremost, from environmental and domestic water requirements and considerations. Even more, if possible, provide an environment in which water is more bountiful. Some novel technologies emerge as these concepts are pursued, as discussed later, to enable us to meet the challenge.

5.5.2 Agriculture

Although employment is gradually developing in other sectors, the bulk of the population, about 75% currently, is engaged in agriculture and it is the dominant sector of the economy. Rapid modernizaton of agriculture is the highest priority activity for development, eradication of poverty, and environmental conservation. Conjunctively, we have also to consider how to manage movement of people out of the agricultural sector, but not away from their natural habitat in villages.

Little has been achieved so far in terms of modernizing the agriculture sector of India. As one reviews the world agriculture scene, two world systems glaringly stand out: (1) the highly efficient agriculture of the developed countries with high productive capacity and high output per worker and (2) the inefficient and low productivity agriculture of developing countries. The gap between the two kinds of agriculture is immense. The per capita agricultural production (dollars) in the former was 1660 against 63 in the latter in 1980 (about 26 times higher), and was estimated to widen in the future (Todaro 1989). The productivity gap had widened to about 50 to 1 by 2000 (Todaro and Smith 2003). The irrigation performance is one of the poorest in terms of agricultural yields also. In a recent study of yields in 93 developing countries, excluding China (for valid reasons as will be brought out later), the average yield was 1951 kg/ha, whereas that of India was only 1600 kg/ha. There are complex socioeconomic–environmental factors for this huge variation, but availability of water is an important factor. The yield in China, which has about the same irrigation characteristics as India and about the same irrigation intensity and cultivated land area, is about 4329 kg/ha. The yields in industrialized countries are around 6000 kg/ha (World Resources Institute 1996).

Kharif irrigation is still not provided, except for some minor developments. Large-scale water withdrawals from rivers and polluted inflows have turned rivers into fetid sewers over large stretches, in view of the increasing urbanization and industrialization, leading to a serious problem of water supply to urban habitats besides the serious environmental degradation. There is the increasing problem of groundwater overexploitation and its pollution. The story has just begun.

Modernization of agriculture is tremendously important for economic development and also for ensuring environmental sustainability, as agriculture is the largest user of land and water. Referring to water, about 80% of it is currently used in agriculture. Reference to yields achieved in modern agriculture, which are almost three times those currently achieved in India, brings out that there is tremendous scope for modernization, leading to the possibilty of giving land and water back to nature (Waggoner 1997).

5.5.3 Modernization of Industry

Leapfrogging to the latest technology and management is essential in view of the economic competition of globalization. With the tremendous pressure on the environment on account of a very large-scale increase in the industrial sector, it is not only the issue of managing the return flows, but also the entire set of operations has to be managed to ensure that the transformation of the raw material is economically productive and environmentally sustainable. This has been called industrial ecology. It has been spelled out in detail (Frosch and Gallopoulos 1989). Quantity of water use and quality of the return flows are components of the totality of activity.

5.5.4 Energy

A revolution is required in energy policy and technologies. With the very large-scale increase in use of energy in all sectors, the demands on resource base and adverse impact on the climate are going to be phenomenal. Thus, it has to be developed as a part of the rapid development of the society, in which the revolution in the sector has to be embedded. An issue of particular importance is development of divisible energy. This is most important for transportation but is also needed for the development of the groundwater. Much work is going on in the sector and hope is generated.

Next to the agricultural sector, the energy sector is the second largest user of water. Thus, the issue is to develop the entire sector so that output is maximized, and adverse impacts are acceptable. Opportunities exist for meeting the challenge, but urgent committed research is required.

5.6 Some Conceptual Revolutions about Management of Water

We are proposing several conceptual revolutions regarding management of water. First, water is a part of the environment and is one of the components. Therefore, we will have to look on the management of the environment in total. As we start looking at water, the perspective all throughout is to consider water as part of the environment, with water as an important dynamic component. This central thought has to be the foundation of all thinking. Land and water are the two most important components, and we consider them conjunctively all throughout.

Second, although we may be talking in terms of river flows, our perspective all the time is the total drainage basin and the land–water system. At the operational level, we are always talking of the management of the smallest unit, that is, the watershed. River flows are just the reflection of the continuing response of the land system to hydrologic excitements.

Third, there is a life cycle of management of water. It has four distinct components. (1) It is diverted from the natural hydrological cycle. (2) It is conveyed and delivered for use in human activities. (3) It is used in human activities. (4) It is returned to the environment. It has to be managed over the entire life cycle in each component, increasingly efficiently, but integrally over the life cycle.

Fourth, the issue is not merely passively managing the environment and its vectors such as water but management of the forcing component—the societal objectives and activities.

Finally, with the advances in science and technology, we undertake interactive participatory management, unlike the traditional engineering.

5.7 Some Unique Indian Environment Characteristic Features

There are two features of the Indian scene, one being physiographic and the other hydrologic, which characteristically differentiates it. On account of the geological history, the Indo-Gangetic basin has a unique physiography. It has the world's youngest and very steep mountains, Himalayas, followed by very flat alluvial plains, and then by the world's oldest plateau–mountains. This applies to the entire northern India but is brought out most impressively in the Ganga–Brahmaputra–Meghna (GBM) basin, as shown in Figure 5.2. Hydrologically, the natural water supply—the precipitation—is highly

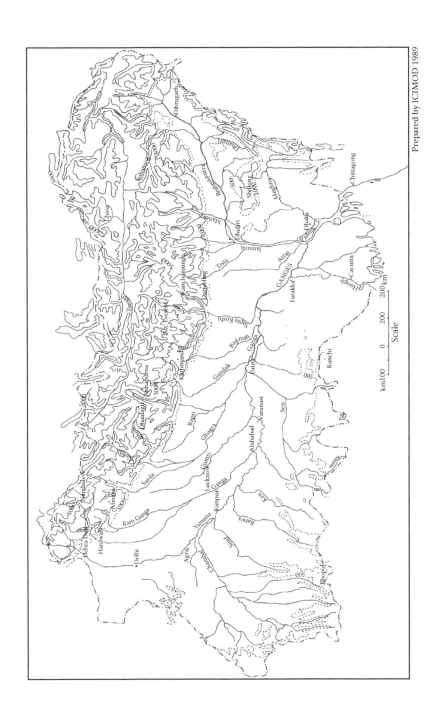

FIGURE 5.2
GBM basin—general features. (From Bruizneel 1989.)

concentrated over a few monsoon months, as shown in Figure 5.1. This leads to the characteristic feature of very concentrated river flows flowing down the Himalayas, meandering over the plains, bound on the south by the plateau formations, from which also some rivers flow down to the GBM basin. As stated earlier, it is distinctly different from the characteristic features of the water supply in Western Europe and, accordingly, significantly shaped the history of the two regions.

The two features, hydrologic and geological, lead to a very concentrated and uncertain availability of water, which in turn leads to the world's heaviest floods during monsoons and considerable scarcity during winters. They also bring huge amounts of sediment and, therefore, by virtue of intense flow concentration, sediment load and alluvial flood plains meander. Kosi has meandered over hundreds of miles. As populations increase and forests gradually disappear, all these forces have fewer restrictions.

5.8 Proposed Creative Approach and Some Novel Technologies and Advances

Two central creative approaches in our thinking, which led to several novel technologies and revolutionizing the availability of water, are as follows: (1) We focus on developing the high monsoon flows, which account for most of the water. Dams, of course, have been proposed for long, but the scope for storing waters through dams in India is very limited on account of the physiographic conditions. We develop some novel ways of managing the monsoon waters, and (2) we try to put the water in the hands of the farmer, instead of his being dependent on the official agencies that have been performing this task, as far as the surface flows were concerned.

If one reviews the physiographic and hydrologic characteristics of India creatively and try to break the bonds of historicity of canal irrigation, which bind even the present policy and water resources development, and undertake the development of water in terms of the natural environmental characteristics, several novel technologies emerge, which serve these objectives nicely. We consider each of the characteristic environmental regions one by one, as they are distinctly different.

As Nehru said, "The Ganga above all, is the river of India which has held India's heart captive and drawn uncounted millions to her banks since the dawn of history. The story of the Ganga from her source to the sea, from old times to new, is the story of India's civilization and culture" (Nehru 1947). We focus on the GBM basin, but these technologies apply to entire India, considered integrally. The GBM basin is the focal region, but the technologies are equally applicable in the Indus basin, though with much limited potential, as discussed in Chapter 6. They are developed while considering all of India as one unit.

5.8.1 Novel Technology 1—Chaturvedi Water Power Machine and Management in Himalayan Region

As we review the characteristics of water resources in the Himalayas, as brought out in Figures 5.1 through 5.4, an opportunity for revolutionizing the water resources development and a novel technology presents itself.

First, focusing in the hill areas, we stress the watershed management in operational terms to conserve the environment, particularly in view of the energy of water on account of the potential kinetic energy. We try to conserve the soil, meet the domestic requirements, and utilize the low flows of the rivers for irrigating the adjoining lands as the layout of the land permits.[1] Next, we try to develop the water potential. Storages are not possible in the head reaches in the Himalayas on account of the steep slope, and the hydro energy potential is usually developed through run-of-river hydroelectric schemes, in the first instance. The plan of one of the sub-basins of the GBM basin, the Yamuna–Ganga–Ramganga system, with current schemes of developing them, as shown in Figure 5.5, brings out the characteristic. Similar developments on all the GBM tributaries are proposed to be undertaken, as shown in Figure 5.6. All these are illustrations of the characteristic features of all the Himalayan Rivers, whether originating in India or Nepal Himalayas, and their development. An important point to emphasize at this juncture is that, in addition to the proposed developments on major rivers, we are proposing storages on the smallest elements as socioeconomically feasible.

The run-of-river projects are currently designed to generate hydroenergy on a 90% reliability basis. The focus, automatically, becomes the low flows. People's protests have started, and the government had to halt the development of these projects. Some attempts have been made by the author to utilize the higher potential of hydroenergy.[2] We are proposing to revolutionize the scene.

On the storage projects, the energy of the flowing water is now generally developed, internationally, on a diurnal pumped storage basis because the peaking power is almost five to nine times more valuable than the firm power, depending on system characteristics. Under the arrangement, the necessary balancing storage structure is located at the toe of the dam to store the daily diurnal water balance (Hall and Dracup 1970). This change has also been introduced by the author in India on a major dam.[3] There is, however, a serious limitation about storages in the Himalayan region in India. The storage projects can store only about 20%–30% of the runoff and, therefore, a vast proportion of the energy cannot be stored and utilized.

How do we store this energy, repeat energy, of the flowing waters in the Himalayas, all along as the waters flow down? Reference to Figure 5.1 will show the tremendous hydroenergy potential of the Himalayan Rivers.

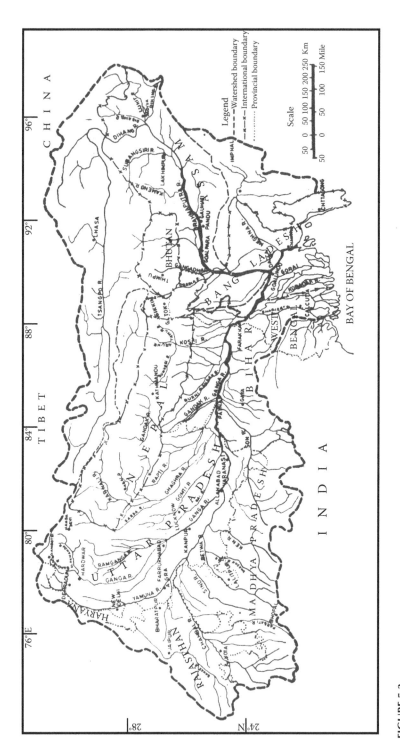

FIGURE 5.3
Layout of the GBM basin.

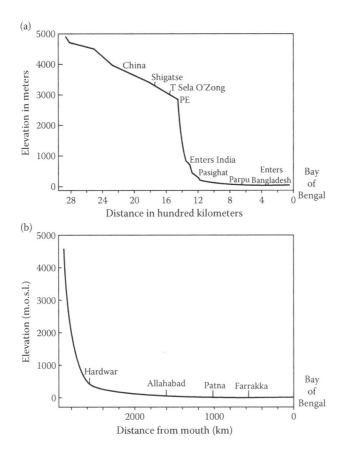

FIGURE 5.4
Longitudinal section of (a) Brahmaputra and (b) Ganga.

Taking the hydrologic and physiographic characteristics of India into consideration, we are proposing a unique novel scheme. It is interspatial or inter-river basin and intertemporal pumped storage of hydroenergy called Chaturvedi Water Power Machine.

It is well known that the Central Highland region of India is one of the oldest geological formations, and the Himalayan region is one of the youngest of the world. The result is excellent storage potential in the Central Highland projects and pitiful storage in the Himalayan storages. This is shown quantitatively in Table 5.1, which gives the height and storage characteristics of some of the Himalayan and Central highland projects.

The proposed plan of development emerges immediately: Generate hydroelectric energy on run-of-river above and beyond the minimum flows conserved from environmental considerations, use it to pump up water and store the energy/water in the central highland projects, and release water on pumped storage basis, later, around the year. Similarly, for the storage

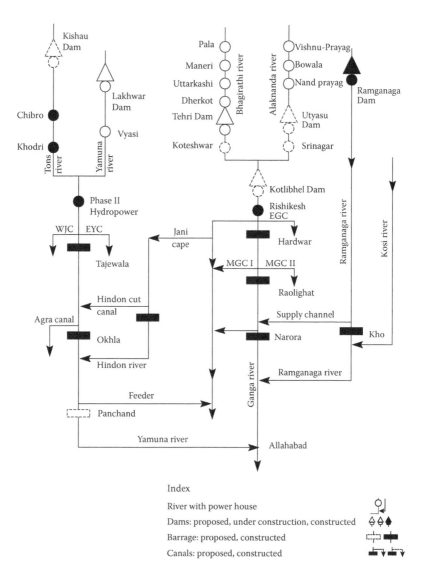

FIGURE 5.5
Yamuna–Ganga–Ramganga system.

projects, one should develop the energy of all the flows, including the proportion that cannot be stored, and store it as proposed.

We examine each feature of the proposal in some detail. The run-of-river schemes are currently designed to provide hydroelectric energy on a daily 90% reliability basis. In contrast to the present practice, we are enhancing the potential of each run-of-river scheme in two ways: (1) increasing the potential and (2) developing it on pumped storage principle as part of the

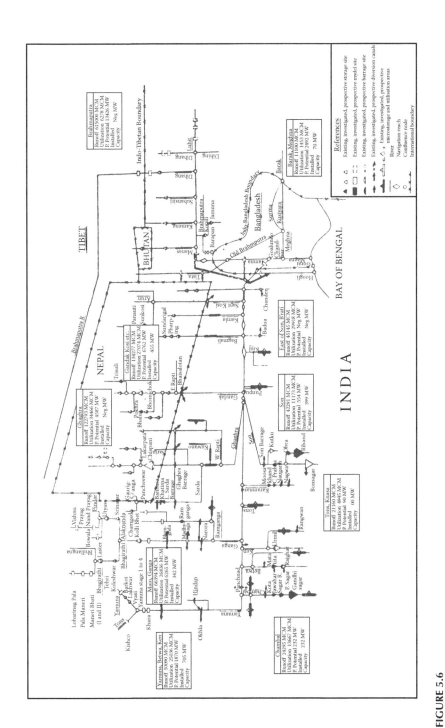

FIGURE 5.6
Macro- and micro-level projects for irrigation and hydroenergy development in Ganga–Brahmaputra–Meghna river system (schematic).

TABLE 5.1

Salient Characteristics of Storage Projects in the Himalayan and Vindhyan Regions[4]

Sl. No.	River	Dam	Height (m)	Storage (MCM[a]) Gross	Live
1.	Ganga	Tehri	260.5	3540	2615
2.	Rihand (Son)	Rihand	93.5	10,800	8967
3.	Chambal	Ranapratap	54.0	2900	1573

Source: Central Water Power Commission, *Storages in River Basins of India*, Government of India, New Delhi, 1997.

[a] MCM = million cubic meters.

proposed intertemporal interspatial complex, in which it will be designed on about a 20% reliability basis instead of the current 90% reliability. The implications of this tremendous increase are brought out by reference to Figure 5.1.

Two aspects of the hydroelectric projects may be mentioned. First, the cost of the run-of-river schemes does not increase very acutely with installed capacity.[5] Second, the design of a hydroelectric project is not a simple straightforward matter. A hydroelectric project involves large initial capital outlays. Almost the entire cost of the project is in terms of its construction. It has to be designed in terms of its long-term analysis of a complex system yet to unfold. It is, therefore, essential that the system is well analyzed as a long-term evolving system with several varying and uncertain features. However, the science of systems planning is well advanced to carry out the study. This subject will be discussed in Chapter 6 as the planning of the proposed technologies is taken up.

The aspect of the proposal needs to be investigated thoroughly before it can be considered as proven for development. The waters of the Himalayan Rivers will be heavily loaded with sediment. Currently, the practice is to clear the sediment load through storage of water in stilling tanks. This may not be possible on account of the large volumes of water. The solution appears to be as follows: (1) provide fine screens on a rotating frame at the tunnel head before the power chamber, the screens being washed off as they move away from the tunnel head, and (2) develop turbine blades that are highly corrosion resistant. The subject needs further research.

Another exciting possibility emerges. The layout of the interbasin transfer proposed by Dr. K.L. Rao is shown in Figure 5.7. Unlike the current proposal of interlinking of rivers of India (which is based on the principle of transfer by gravity as far as possible, the feature assuming economic viability, which may not be valid as discussed in Chapter 4), Dr. K.L. Rao proposed pumping of the Himalayan waters after the major northern tributaries Ghagra and Gandak had met Ganga, just upstream of Sone near Patna up river Sone, which is the major Ganga tributary from the south. It was planned to be transferred south after crossing the Vindhyas. It was proposed even then

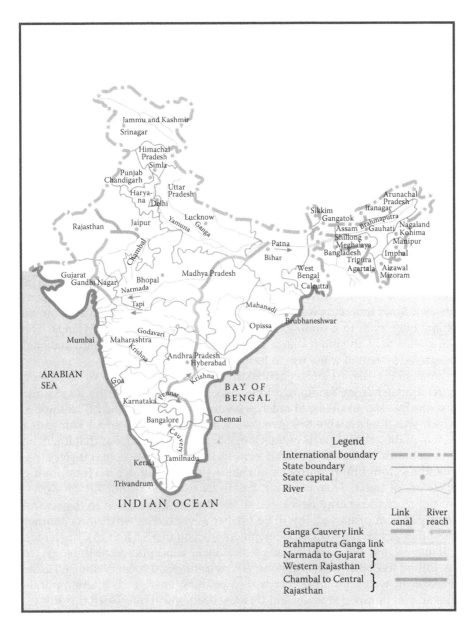

FIGURE 5.7

National water grid. (From National Commission for Integrated Water Resources Development (NCIWRD), Report, Vol. I, *Integrated Water Resources Development—A Plan for Action*, Ministry of Water Resources, Government of India, New Delhi, 1999. With permission.)

that, besides pumping water, the proposed approach should be adopted (Chaturvedi 1973).[6]

Reference to Dr. Rao's proposal is made only to bring out that transfer to the southern peninsular area may be undertaken by negotiating the Vindhyan divide through pumping waters up. The route and capacities have to be decided based on detailed studies in terms of economic development, sustainability, and desirable balanced regional development. It is not proposed that the same route be necessarily followed. We also make a small improvement by bringing Kosi water also to meet Ganga upstream at the point of transfer to south via Sone, or transfer Ganga water to Sone after Kosi meets Ganga. A major advancement is that we now convey water and energy. We can develop additional energy en route, adding the possibility of diurnal pumped storage, beyond the Central highlands, as the water is transferred south. We can discharge part of the water/energy back in the GBM basin, if so desired and found appropriate, but this does not appear to be the preferred policy.

Rao envisaged only minor transfer from Brahmaputra. As he noted, "The waters of the Brahmaputra, as is clear, cannot be used in any appreciable quantity as it flows in a narrow valley hardly 80 km wide. The only way it can be of any help, even to a small extent, is by diversion to assist the Ganga. The Ganga has to be the main source of surplus water for planning any national integrated system of water use in India" (Rao 1975, p. 219).

We endorse Rao's views, particularly after the experience of designing water transfer from Ramganga to Ganga, in the context of the Ramganga project. Transfer in the head reaches involves negotiating the hill channels, which was a formidable task even while transferring the small Ramganga waters in the western parts of the Ganga basin. It is considered that it will be uneconomical and unsafe, even if technically accomplished, to transfer large quantities from Brahmaputra, as currently proposed under the Interlinking of Rivers of India. This, however, has to be examined in detail. However, we can store the energy developed in the Brahmaputra basin in the central energy system proposed. It will be further advanced as we discuss another revolutionary technological development in Section 5.8.5 and in Chapter 6.

It may be noted that with these revolutionary concepts and technology, we are killing five birds with one stone: (1) environmental conservation, (2) large increase in water, (3) large increase in peaking energy, (4) intertemporal–interspatial pumped storage, and (5) most economical interbasin transfer.

5.8.2 Novel Technology 2—Chaturvedi Ganga Water Machine—Management of Water in the Alluvial Indo-Gangetic Basin

The characteristics of the water availability in India, as brought out in Figure 5.1, show that the dominant proportion of the water, almost 80%–90%, is during the monsoons. The water flows down on an alluvial plain, on surface and numerous small rivulets, which have been carved out on it. The alluvial plains are one of the world's best groundwater transmitting media leading to one of the richest dynamic and static groundwater storages in the world.

However, the physiographic–hydrologic characteristics also lead to some serious problems. The heavy precipitation during the monsoons, concentrated over limited time, often results in abundance of water over the alluvial plains, further accentuated by concentrated flows from the steep Himalayas, which has limited time to flow down the rivulets and rivers in their alluvial plains regime in view of very easy gradients and the topographic characteristics. Serious flooding takes place. On the other hand, often the rains fail and droughts and famines result.

There is another characteristic of the Indo-Gangetic plains adding to the woes of monsoon rainwater discharge. The precipitation increases as one moves east, being one of the highest in the world in lower reaches. This feature minimizes easy drain-off and further increases the challenge of managing water as one moves downstream in the GBM basin.

A historic dilemma results. A farmer, depending on rainfall during the monsoon period, the Kharif season, is faced, on one hand, with uncertainty of water availability in some years and, on the other, abundance of water and floods in some others. During the nonmonsoon period, the Rabi season, he is acutely short of water every year.

It presents a challenge. Can we store the abundant monsoon waters in the flat Indo-Gangetic plains as different from storing them in the Himalayan region, thereby resolving the water supply problem for irrigation on one hand and minimizing the flooding problem on the other? Studies carried out indicated that it is possible (Revelle and Herman 1972; Lakshminarayana and Revelle 1975; Revelle and Lakshminarayana 1975; Chaturvedi et al. 1975; Srivastava 1976; Chaturvedi and Srivastava 1979).[7]

The basic scientific principles of groundwater flow clearly show that it depends on three factors: (1) the differential between the water head at inlet and outlet or on the top surface and the groundwater level—the greater the gradient, the higher the flow; (2) the time this activity takes place; and (3) the characteristics of the transmitting medium. Ganga alluvial plains have excellent physical characteristics for groundwater flow. However, a caveat may be added. This is under saturated flow conditions. The entire groundwater flow regime will not be offering saturated flow characteristics.

The operational policy of managing the waters of the Ganga alluvial plains follows. In complete contrast to the current policy of water resources development, the first priority is management of the monsoon period waters over the river basin, instead of the low period, as practiced currently. Try to increase the availability, detention, and time over the river basin. The objectives are (1) providing assured Kharif irrigation, (2) increased groundwater recharge, and (3) management from the flooding perspective. This is achieved by (1) distributing the Himalayan river inflows from the main rivers into the numerous drains, (2) managing the flows in the alluvial plains and in their numerous drains (*nallahs* in local parlance), (3) increasing their storage time by constructing small series of check dams, (4) storing the water in natural storages, (5) increasing detention time on the fields, and so on. An interesting technology of storage, collapsible plastic dams, has been reported in literature and was worked out by the author.[8] The channel systems are treated as part of the drainage basin, fed by it through precipitation and feeding it for recharge. We re-emphasize that we are looking in appropriate spatial terms, that is, in terms of economical management of the land–water system and not on individual channels.

Conjunctively, we increase the differential head for groundwater recharge by lowering the groundwater table at the beginning of the monsoon period by undertaking irrigation extensively in the rabi period through groundwater instead of the current policy of diverting all the low river flows, after the monsoons, to give an apology of irrigation at the mercy of the officials in the colonial tradition. The low river flows after monsoons should be conserved first and foremost for domestic water requirements of the exploding population and environmental conservation, and only the residual, if any, should be used for irrigation, if needed at all. This will ensure adequate, timely, and reliable water supplies, leading to the high yields. Conjunctively, the water table is planned to be lowered to designed levels. It may be emphasized that we are not focusing on the individual small channels but on the entire alluvial plain system, though in the last analysis, the interaction at the rural level will be through these small streams. The two perspectives are developed conjunctively to lead to the water management policy.

We add another novel feature to the proposed policy. We have developed a new type of small tube well, which is reversibe, called the Chaturvedi Reversible Pump (CRP).[9] This will be used, principally, for groundwater development during rabi. However, surface water supplies can be supplemented by groundwater supplies for irrigation, or the surface waters can be pumped down to mitigate flooding and contribute to recharging of the groundwater, during the monsoon period, as the case may be.

It may be noted that, theoretically, facilitating downward flow of water will not need any energy consumption. On the contrary, it can be used for energy generation. This may not be always possible in real life as the starting of flow may be hampered because of the siltage at the entry, but once started, the flow will take place on its own, though it can and may have to be enhanced. For achieving rapid rate of recharge, forced pumping may, however, have to be adopted. However, all this has to be worked out.

To sum up, the low flows, during the rabi period, shall be conserved for providing drinking water and environmental conservation, in the first instance, and they will be diverted for irrigation only if found surplus. The development in the rabi is planned based on lowering the groundwater up to predesigned levels, providing rabi irrigation, so that the recharge during the Kharif period takes place. The Kharif irrigation is provided through natural precipitation, supported by Kharif surface irrigation. Besides management of surface flows for this purpose, operations are supported by CRP, which provides surface water for irrigation, if need be, or pumps down water for storage and flood mitigation, as the need arises. The important perspectives are making water available at the command of the farmer and management of water integrally, spatially, and temporally. The entire scheme is called Chaturvedi Water Machine, for the sake of easy reference. This scheme has been well tested through mathematical modeling (Srivastava 1976; Chaturvedi and Srivastava 1975). Proposals were made for undertaking field level experiments by the then Minister of Irrigation, Dr. K. L. Rao, and later again at the highest level, but they have not been implemented.[10]

Another point to be emphasized is that although the idea looks simple, and the local people's participation is important and should be encouraged, the activities must be guided by knowledgeable persons. The subject is sometimes covered simplistically under "local water resources development and management" (NCIWRD 1999). The example of Lava Ka Baas, with which the author was associated, and which failed after the first monsoons, is exemplary.[11] Similarly, "rain water harvesting" is much publicized. Even urban rainwater harvesting is encouraged by some agencies and even official agencies. This poses problems because of the serious water quality implications. Management of groundwater is a very complex issue and should be undertaken only under the guidance of knowledgeable persons.

5.8.3 Novel Technology 3—Development of Artesian Waters of the GBM Basin

One of the geological characteristics of the Himalayas, resulting from the history of their creation, is the formation of two characteristic land forms, called Bhabhar and Terai, at their feet before the flat alluvial plains are laid out. Bhabhar is a level surface zone at the foothills of the Himalayas, about 30 km wide where the Himalayan rivers and streams disappear under boulders and gravels due to the porous soil and subsoil composition of Bhabhar. As a result, the underground water level is very deep in this region.

During the course of investigations in the GBM basin by a team of Indian and foreign scientists, in the context of oil investigations, it was established that the region is well served with artesian waters because of these characteristics. It had been proposed that the artesian waters could profitably be used for water supplies (Jones 1985). The subject was well publicized by the World Bank.[12] However, for some reasons, it is not mentioned at all now. The subject should be studied, as these are very valuable resource and energy sources.

5.8.4 Novel Technolgy 4—Integrating Static Groundwater

Besides the dynamic groundwater resource, there is a very large static fresh groundwater resource in the Indo-Ganga–Brahmaputra basin. The figures of the two, static and dynamic, for the three river basins are 1338.2/26.49,

7825.3/170.99, and 917.2/26.5 km³, respectively (NCIWRD 1999). A dynamic groundwater development policy could be adopted, which may overcome the problems being encountered on account of delays in the storage projects. It is emphasized that it is not proposed that static groundwater be mined. It is only supposed to establish and assure the final water resources development activities as an interim measure, if need be.

5.8.5 Novel Technology 5—Integrated Water and Energy Systems Management

Conjunctive surface and groundwater use is well known. Similarly, conjunctive thermal and hydroenergy advantages are known. We propose conjunctive use of all the four, which will lead to large economies of production and yield. The region offers some unique and rich conjunctive surface, groundwater, and energy management opportunities in view of some significant characteristics. The relative surface and groundwater potential is almost of the same order of magnitude and spatially and temporally very varied. The hydropotential is very large, and thermal potential is also of a large order of magnitude and temporally very complementary. The pumped storage has valuable and large intertemporal and interspatial potential in contrast to

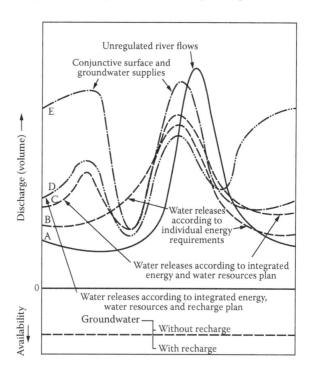

FIGURE 5.8
Conjunctive surface and groundwater development for integrated water and energy planning.

the conventional diurnal characteristic of limited potential. Considering all these factors together, significant advantages can be obtained if integrated management of water and energy is undertaken. It was in this context that the use of static groundwater was mentioned.

A schematic perspective has been developed as shown in Figure 5.8, but it has to be developed in detail. Planning is complex, but modern advances in systems planning enable it to be worked out. Some studies have been undertaken, as discussed later.

5.9 Estimate of Water Availability Revolution

A brief assessment of water availability, in view of the proposed advances, may be given to give an idea of the magnitude of increase. The estimates are tentative because the projects have to be developed in terms of the economic viability. However, it is considered that the proposed estimates are realistic.

According to the current official estimates, mean surface flow, utilizable surface flow, replenishable groundwater, and total utilizable water are 1952.87, 690.32, 342.43, and 1032.75 km^3, respectively (Table 5.2). It was argued that the utilizable surface flows of the peninsular rivers can be increased to their full mean value, without any novel technology, increasing the total of utilizable surface flows to 776.94 km^3 (Chaturvedi 2001). Similarly, the utilizable groundwater was estimated at 500.43 km^3, in accordance to the Central Ground Water Board (CGWB) recommendation, making the total of utilizable water resources 1276.47 km^3, in contrast to the current official estimate of 1032.75 km^3, according to the earlier estimate called Chaturvedi 1.

We are vastly increasing these estimates through the proposed novel technologies. We are arguing that through the proposed novel technologies, almost all 1275.46 km^3 of surface flows of the rivers of India, excluding the Brahmaputra and Meghna, can be utilized. Adding the estimated utilizable figures of these two rivers of 24.00 km^3, we obtain a figure of utilizable surface waters as 1299.46 km^3, in contrast to the current official estimate of 690.32 km^3, an increased figure of almost 1.9 times the current estimates. In addition, there is an increase in groundwater availability through the proposed Chaturvedi Ganga Water Machine (CGWM). This will increase the groundwater availability much above the currently total estimated availability by the CGWB of 500.43 km^3, but because estimates are difficult to make, we have limited ourselves to the CGWB figure of 500.43 km^3. We come to a total figure of utilizable waters as 1799.89 km^3. Thus, there is an increase of about 88% in utilizable surface waters, 46% in groundwater, and 74% in total utilizable waters. The transfer to the peninsular region of the estimated currently unutilized surface waters of Ganga, 275.02 km^3, represents a vast increase from the current estimated value of the utilizable waters of

TABLE 5.2

Estimates of Water Availability (km³)

| Sl. No. | Agency | Surface Water | | Groundwater | Total | Peninsular[a] |
		Mean	Utilizable			Surface Water
1.	NCIWRD	1952.87	690.32	342.43	1032.75	229.17
2.	Chaturvedi 1[b]	1952.87	776.94	500.43	1276.47	298.90
3.	Chaturvedi 2[c]	1952.87	1299.46	500.53[d]	1799.89	573.92
4.	Increase[e]		88%	46%	74%	150%

[a] Peninsular rivers considered are Godavari, Krishna, Pennar, Cauvery, Tapi, Narmada, Mahi, Sabarmati, and west-flowing rivers of Kachchh and Saurashtra including Luni, to which the Ganga waters could be added.
[b] Chaturvedi 1 estimates relate to increase based on conventional advances.
[c] Chaturvedi 2 estimates relate to increase through novel technologies.
[d] As stated, the estimates are tentative.
[e] The increased estimate is based on the novel technologies.

the peninsular rivers Godavari, Krishna, Pennar, Cauvery, Tapi, Narmada, Mahi, and Sabarmati, and west-flowing rivers of Kachchh and Saurashtra including Luni, to which these could be added, estimated as 229.17 km³. It was shown that this could be increased to 298.90 km³, even by some conventional advances (Chaturvedi 2001). Further adding the figure of 275.02 km³, we get the figure of 573.92 km³, which is an increase by about 150%.

The scene is summarized in Table 5.2.

We are also arguing that the figures of utilization, particularly in the agricultural sector, which currently accounts to about 80% of the utilized waters, can be substantially reduced. Thus, a revolution can truly be brought about in the management of India's waters. The central point, however, is creativity, modernization, and, above all, implementation. It must, however, be clarified at the outset that achievement of these figures depends, first and foremost, on the state of the social system and is not a matter of technology alone.

5.10 Further Advances in Water Management

We have so far focused on development of water. We have to consider the totality of the activities in the context of the water cycle, which consists of four distinct phases of development, transmission, use, and management of return flows.

Advancing to use or demand, the first one is for personal consumption and needs. This requires a small fraction and can be easily met from the vast amounts developed. It has been demonstrated that considerable advances in transmission can be made, where a large proportion is lost. In the rural sector, the transmission does not even exist. There is, thus, a considerable scope for advancement over this phase. However, it has to be addressed in terms of the complex from which it is emanating—urban and rural systems.

Focusing on agriculture, which has the overwhelmingly large demand on water, modernization of the irrigation system, which accounts for all the three phases, has been well emphasized in literature and discussed in NCIWRD (1999). There is scope for very large advancements. For example, it has been demonstrated that even if we manage land leveling and application properly, still limiting to simple current technology that can easily be practiced by the farmers, there are considerable gains in yield and saving in water (Khepar and Chaturvedi 1979; Khepar 1980). Much higher gains can be achieved, on both these counts, with increasing advances of technology of application, as demonstrated in California and Israel. For example, the productivity of water is a function of the stage of growth of a plant. Therefore, considerable economies in water use can be made by judiciously deciding the timing of applying water (Arya 1980; Arya and Chaturvedi 1981). As another example of the importance of management of return flows in this context, it may be pointed out that water introduces biochemical changes in the land–water system. The salinization of the Indus basin is a classic example of the disastrous impacts of the neglect of this phase of activity. The challenge of management of the arsenic, highlighted in Bangladesh, is another frightening example of an even more complex issue of groundwater management, as discussed in Chapter 6. The simple conclusion is that we have to manage the totality of the land–water system and that too taking into account the historicity of the system. The focus may change to land, and watershed management may become the perspective of development at some stages and water at the other. Again, it is embedded in the societal system.

Similarly, water in the industrial system has to be developed in the context of modernizing the entire industrial activity cycle, called industrial ecology (Chaturvedi 2011a).

With the changing environmental complex and socioeconomic system, the flow in rivers has to be managed very scientifically. CGWM will lead to enhanced low flows, which have to be adequately used to meet the domestic and industrial requirements. However, there will be vastly increased and polluted return flows as economic development takes place. Therefore, the focus of development and management may change in the context of undertaking the environmental conservation, managing both the enhanced flows as well as the return flows. The return flows are part of the complex land–water dynamics of the river basin. They are not merely direct inputs to surface flows but have a complex nature. The rivers do not get water just from the mountains. There is an increasing input from surface flows and groundwater interaction as it flows along. The river is just one phase of the land–water system over the basin. As chemical fertilizers are fed to the land, there will be increasing polluted inputs from the groundwater. The interaction between river and the adjoining groundwater system is dynamic, changing over the course, each becoming a supplier or recipient, as the case may be (Prasad 1981). Management of the return flow is extremely important as they cannot be managed just by dilution, as is the current thinking.

A land–water system is a complex system with varying characteristics from the mountains, the plains, and the delta as its well-defined features. It varies over each of these. It has also to embrace the land's end and has to be managed in an integrated manner over the totality.

5.11 Environmental Systems Management

The review of the story so far will clearly bring out that water resources development cannot be undertaken in simple engineering project terms and by considering development of the water as a resource, water and energy sector, or environmental impact, in isolation, as is being done currently. We have to consider integrated surface water, groundwater and energy management, and enironmental management. It has to be undertaken integrally, spatially, and temporally using several spatial and temporal scales. In the first instance, we focus on the enginering aspects and consider some issues.

If releases are made on a variable irrigation pattern depending on water requirements of the crops, the hydroelectric generation of storage projects will vary over the year. In India, because hydroelectric generation so far has been small, and the irrigation requirements from surface water are dominant, the releases have been planned to be made according to irrigation requirements, and a hydroelectric cushion in the total storage is generally provided. However, this will not be the best practice as hydroelectric development increases and groundwater compliments surface water, and at the same time demands energy. Further complication arises from two considerations. What is the optimal cropping pattern and what are the corresponding optimal irrigation release implications? The picture gets further complicated because the water supplies can be met both from surface waters and groundwater, with the former contributing energy in varying proportions to discharge as the head changes and the latter requiring energy for production. But it has to be worked out in the context of the foregoing conjunctive surface water and energy characteristics of production and demand. Further complication arises as hydroelectric energy and thermal energy are planned conjunctively, with varying opportunities for run-of-river and storage projects, and conjunctive interlinkage with irrigation demand and energy generation for groundwater development and through surface water release characteristics and potential. Diurnal pumped storage is added to the possibilities as one immediate response. What is the picture as such a large system of any river basin is developed at a large scale, leading to complex dynamic system management whose trajectory itself is dependent on the emerging variable economics of the system, compounded by real-life uncertainies of implementation?

A complex picture of conjunctive surface water, groundwater, and hydro-electric generation, embedded in the energy system, emerges as one contemplates the development of the GBM basin. The system is developing. The region offers some unique and rich conjunctive surface, groundwater, and energy management opportunities in view of some significant characteristics, as follows. The relative surface and groundwater potential is almost of the same order of magnitude and spatially and temporally very varied. The hydro and thermal potential is also of comparable order of magnitude and temporally very complementary. The pumped storage has valuable and large intertemporal and interspatial potential in contrast to the conventional diurnal characteristic of limited potential. Considering all these factors together, significant economies can be obtained in the management of water and energy.

5.12 Modernization of Socioeconomic System

The focus of the developmental policy of water has been on water as a physical entity or as a resource, and the central theme has been the enhancement of its availability, particularly for the agricultural sector. The logical question arises in that the economic sector should also be developed conjunctively so that the resource is used more efficiently, leading to greater returns, on one hand, and minimizing the use of resources and consequently adverse impact on the environment, on the other. Therefore, conjunctively with the development of the physical system, the economic system should also be made more efficient. Thus, if the agricultural system is advanced, it will have varied and positive impacts on the physical system. Advancement of the agricultural system, which is the dominant user of water, provides a valuable opportunity. For example, with the new seeds, and the number and timing of irrigation, there are vastly decreased water requirements for the same amount of crop yield, leading to highly reduced impact on the environmental system. Thus, management of the agricultural system is crucially important for the scientific land–water environmental system as it dominantly impacts it. It is important to emphasize that adequate, timely, and assured water supply is a precondition for the revolution of the agricultural sector.

Extending to the other sector of socioeconomic activity, the habitat and industrial sector, the impacts, again, are significantly dependent on how the sector activities are undertaken. This is a very important issue because it will not be possible to meet the urban demands if the current urban processes, already posing almost unmanageable water management, continue.

Similarly, industrialization is still in a nascent stage. If it is undertaken scientifically, the resource demands can be reduced, and, more importantly, the adverse environmental impacts can be reduced considerably.

5.13 Policy Planning—Societal Environmental Systems Management

The subject of scientific management of water has been advanced, recently, to cover these issues. It may, therefore, be reviewed in this perspective to help in its adoption, as it is lacking generally and particularly in India.

Advancement of the management of water has been a pioneering activity with increasing advancement of science and technology. Fusion of science and technology increasingly occurred over time. Because engineering involves economics, attempts at undertaking the activities optimally in economic terms have always been made. However, there was a revolutionary advancement when scientific engineering–economic management, backed by the emerging science of systems analysis, emerged in the late twentieth century. This found ready preliminary application in water area through the works of some scientists such as Mass et al. (1962) from Harvard, Hall and Dracup (1970), and many others, leading to the science of water resources systems planning. The latest state in the subject has been brought out by Chaturvedi (1987).

With the recent emphasis on sustainable development, attempts to merge the issues of engineering and environmental management, leading to the current concept of Integrated Water Resources Management (IWRM), have been made. However, under IWRM, the concern remains at emphasizing that the water management activities be also evaluated and guided from the concept of the environmental sustainability.

Further advances have been made. Because water is a component of the environment, and environment and society are interlinked, the concern should be in terms of the entire societal–environmental system rather than focusing on environment in isolation. It emphasizes the basic fact that because the management of water is embedded, and is used by the society in its diverse activities, these activities should also be modernized conjunctively, leading to better socioeconomic returns as well as minimizing the adverse environmental impacts. These developments are still in the formative stages (Bossel 1998). This has been formally considered and has been called Societal Environmental Systems Management (SESM) (Chaturvedi 2011a,d).

These ideas are expressed in the following figures. Development and management of waters in the earlier perspective of water resources systems, advanced in terms of IWRM, are shown in Figure 5.9 (Chaturvedi 1987). It

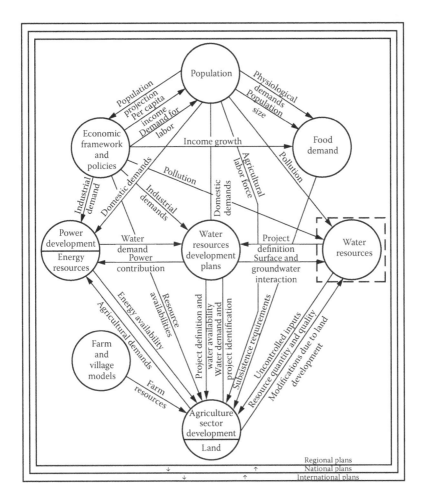

FIGURE 5.9
IWRM concept—resources and policy interactions.

emphasizes that use of water should be considered integrally, in terms of the diverse activities, taking into account their demands and impacts.

With the advancement of concepts and science of engineering planning, the new SESM perspective is shown in Figure 5.10. With the revolutionary advances in information technology, it is possible to develop policies, backed by systems dynamics–based analysis, enabling analysis and experimentation of the proposed policies, taking due account of the complex interlinkages shown in Figure 5.10 (Chaturvedi 2011a). The ideas are backed by systems planning, which can be even transparent, and participatory analysis, thanks to the modern advances in information technology, offering the possibility of on-line interaction with policy planners (Rothmans and de Vries 1997; Hoekstra 1998). Some work has also been done by the author's research

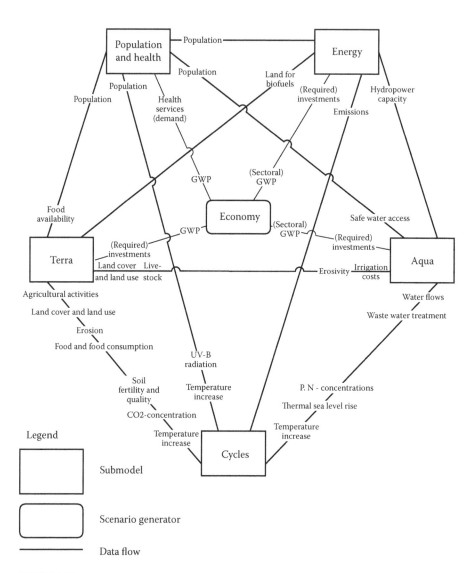

FIGURE 5.10
Societal–environmental systems development. (From Chaturvedi, M.C., *Environmental Systems Management for Sustainable Development* (in press), 2011.)

scholars in the context of India, but much remains to be done (Kothari 2000; Hasan 2005).

Planning and management have to be undertaken at several spatial levels interconnectedly. At one level is the river basin and at another is the smallest scale of rural stream on which storage can be developed. The entire system is shown in Figure 5.11 to emphasize this interconnection.

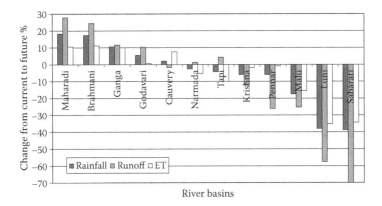

FIGURE 5.11
India's water system—macro to micro.

5.14 Challenge of Climate Change

We have not mentioned climate change so far. Climate change represents a serious additional stress on ecological and socioeconomic systems. It becomes particularly severe in the countries especially of South Asia, in view of the climatic characteristics compounded by the tremendous environmental pressures as a result of rapid urbanization, industrialization, and economic development. With its huge and growing population, an economy that is closely tied to its natural resource base, a snow-covered mountain chain integrally linked with supply of water, and a 7500-km-long densely populated and low-lying coastline, India is considerably vulnerable to the impacts of climate change.

Studies conducted so far have shown that the surface air temperatures in India are going up at the rate of 0.4°C/100 years, particularly during the post-monsoon and winter season. Using models, they predict that mean winter temperatures will increase by as much as 3.2°C in the 2050s and 4.5°C by the 2080s due to greenhouse gases. Summer temperatures will increase by 2.2°C in the 2050s and 3.2°C in the 2080s. Extreme temperatures and heat spells have already become common over northern India, often causing loss of human life.

Climate change also has had an effect on the monsoons. India is heavily dependent on the monsoon to meet its agricultural and water needs, and also for protecting and propagating its rich biodiversity. The pattern of precipitation will change severely spatially and temporally. Studies have shown subtle changes already in the monsoon rain patterns. A decline in summer rainfall by the 2050s, which accounts for almost 70% of the total annual rainfall over India, is crucial to the country's agriculture. Significant change in river flows,

FIGURE 5.12
Change in water balance for climate-scenario control (CTL) and greenhouse gas (GHG) climate scenarios. (From Gosain et al. 2006.)

as shown in Figure 5.12, is predicted (Gosain et al. 2006). Relatively small climatic changes can cause large water resource problems, particularly in arid and semi-arid regions such as northwest India. This will have an impact on agriculture, drinking water, and generation of hydroelectric power.

Apart from monsoon rains, India uses perennial rivers, which originate and depend on glacial melt-water in the Hindu Kush and Himalayan ranges. Because the melting season coincides with the summer monsoon season, any intensification of the monsoon is likely to contribute to flood disasters in the Himalayan catchments. Rising temperatures will also contribute to the raising of snowline, reducing the capacity of this natural reservoir and increasing the risk of flash floods during the wet season. As the estimates are at the highest level by the International Committee of Climate Change, the Himalayan glaciers will be completely lost in about 100 years.

Increased temperatures will impact agricultural production. Higher temperatures reduce the total duration of a crop cycle by inducing early flowering, thus shortening the "grain fill" period. The shorter the crop cycle, the lower the yield per unit area. According to a recent study by an eminent scientist, India will be most severely affected. It is estimated that the output by 2080 will be reduced by about 60.9%, 57.9%, 31.3%, and 36.8% in northwest, northeast, southeast, and southwest, respectively, without taking the carbon fertilization into consideration. This will decrease the estimated reduction, but the estimated impact will remain phenomenally high (Cline 2007).

A trend of sea level rise of 1 cm per decade has been recorded along the Indian coast. Sea level rise as a result of thermal expansion of sea water in the Indian Ocean is expected to be about 25–40 cm by 2050. This could inundate low-lying areas, down coastal marshes and wetlands, erode beaches, exacerbate flooding, and increase the salinity of rivers, bays, and aquifers.

Deltas will be threatened by flooding, erosion, and salt intrusion. Loss of coastal mangroves will have an impact on fisheries. The major delta area of the Ganga, Brahmaputra, and Indus rivers, which have large populations reliant on riverine resources, will be affected by changes in water regimes, salt water intrusions, and land loss. The most severe impact will be experienced in Bangladesh, where about three quarters of the country may be submerged, unless appropriate steps are taken well in time.

Increase in temperatures will result in shifts of lower altitude tropical and subtropical forests to higher altitude temperate forest regions, resulting in the extinction of some temperate vegetation types. Decrease in rainfall and the resultant soil moisture stress could result in drier teak-dominated forests, replacing sal trees in central India. Increased dry spells could also place dry and moist deciduous forests at increased risk from forest fires.

The future scenario in terms of the climate change impacts, according to current studies, is exceedingly disturbing. The implication for management of the water resources to meet them is increasing scientific analysis, above and beyond the foregoing studies in the context of social–environmental concerns, in the first instance. The only saving grace is that the advances that have been proposed will better equip us to meet these additional challenges.

5.15 Conclusion

The foregoing brief presentation will bring out that there are several basic issues that require emphasis and serious consideration. The current official policy needs a thorough revision. India is on the path of rapid development. The future path is not of extrapolation of the path but one of revolution—revolution in concepts and creativity. It will mean interaction and integration of the environment at a very large scale. We have emphasized that the entire spectrum of integrated SESM has to be considered integrally. Furthermore, the approach is not that of making pious pronouncement but backing analysis and creativity with system dynamic studies, as the advancement in the field permits. Besides the basic formulations, some efforts in this context have been made, but they are just the beginning (Chaturvedi 2011a,b,c,d).

Notes

Reference to personal experiences is given just to establish the authenticity of the ideas.

1. The author worked in Himachal Pradesh for about two and a half years as a young engineer walking up from Shimla to the Tibet border, on foot and across the state, several times. Illuminating experience of the challenge of managing the environment in the Himalayas was obtained. The author has also been often invited to Swiss universities to deliver talks on his contributions in the field, which gave illuminating experience of challenges, difficulties, and potential of development.

2. The opportunity of providing the proposed high capacity on run-of-river arose when the author was working as Member, Board of Consultants, Maneri Bhali Hydroelectric Project (MBHP) II. Its capacity had been proposed as 150 MW based on the earlier design of MBHP I, which had the same head and discharge. The capacity has doubled to 300 MW based on the author's recommendations. The basis is advanced more scientifically in the proposed approach.

3. Tehri Dam, the third and the second in the Himalayan region in the GBM basin, has been completed recently. The author was on its Board of Consultants, under the Chairmanship of Dr. A.N. Khosla. The original design of Tehri Dam did not envisage pumped storage. Its design was modernized by the author by providing pumped storage. However, it was on the conventional diurnal basis.

4. The author was on the Board of Consults for the first project, designed the second project, and was a consultant for the third. When the author designed the second, the Rihand Dam, it had one of the world's highest storage, soon to be outclassed by Kariba on Zembasi.

5. Referring to Figure 5.1, this is how Maneri Bhali 2, which has the same discharge and head as Maneri Bhali 1, was designed. It was argued by the author that all the potential that can be generated economically, above and beyond the minimum low flows, should be developed, as the annual temporal variation could be managed through appropriate operation of the power system, particularly because marginal increased cost of the energy with capacity is small in a run-of-river scheme, and the higher capacity may provide, up to a limit, even cheaper energy than as designed on the 90% reliability basis. The attractive feature arises because of the characteristics of run-of-river schemes. It has three components: (1) diversion headworks, (2) powerhouse, and (3) the conveyance tunnel. The cost of diversion works is fixed, independent of the installed capacity.

The cost of the powerhouse varies only marginally with the capacity. The maximum proportionate cost of the system is in the conveyance tunnel. Here, too, the variation with capacity is small because the total cost is the sum of two components: (1) the capital cost of the tunnel and (2) the capitalized value of energy lost in the tunnel. As one increases the tunnel diameter, the capital cost increases, proportionate to the square of the diameter, but the capitalized cost of the energy lost decreases, proportionate to the cube of the diameter. Therefore, the optimal dimension of the tunnel is decided by obtaining the total of the two. Hence, increasing the tunnel diameter does not increase the total unit cost directly, and considerable benefits result in increasing the installed capacity, enabling much higher energy production.

6. The idea of Chaturvedi Water Power Machine was first developed in the context of Dr. K.L. Rao's proposal to pump the Ganga waters up River Sone and transfer to Cauvery (Chaturvedi 1973). It was discusssed with Dr. K.L. Rao, with whom the author was working closely and was appreciated by him because it contributed to his idea of the National Water Grid and Ganga–Cauvery link. However, he left the Government and his ideas have been completely jettisoned by the Interlinking of Rivers proposal.

7. This was referred to in Chapter 4 but is reproduced for the sake of ready reference in the context of the subject under discussion. The challenge of storing the monsoon runoff was raised in an Indo-U.S. Science Academy Seminar in New Delhi in the early 1970s attended by Chaturvedi and Revelle. Several suggestions were made. The author and Revelle pursued the idea and studied to different levels of detail. The scheme was discussed informally by the author with Dr. K.L. Rao, Minister of Irrigation, GOI, with whom he had the privilege of working closely. He approved it and appointed a committee to carry out preliminary tests. This was done and the feasibilty was established. Detailed studies were proposed to be undertaken, but Dr. Rao later left. Revelle and Chaturvedi independently suggested lowering of the groundwater all over the plains, though the approach was slightly different. Revelle gave the name Ganga Water Machine. He accepted Chaturvedi's approach, and this is why it is called Chaturvedi Ganga Water Machine. The two worked closely for the implementation of the proposal. A Workshop of Indian and U.S. scientists was held in 1981 and the subject was discussed in detail. Field visits of the Workshop scientists were organized. The scheme was approved by the Workshop scientists. It was decided that further studies and field tests should be carried out.

8. According to an article in Civil Engineering, sometime in the early 1960s, a paper was published about collapsible rubber dams

developed in Japan. The author was designing the Ramganga Project at that time, which involved construction of a number of diversion structures on the Ramganga Channel. The possibility of collapsible rubber dams was investigated and was found feasible, but it involved considerable foreign exchange, which was scarce at that time. However, it can easily be constructed indigenously now.

9. The scientific basis of the Chaturvedi Reversible Pump has been established by developing a design and having it tested in the laboratory. A prototype is trying to be developed by the author at the Indian Institute of Technology (IIT) Delhi. Application for patent has been made by the IIT Delhi.

10. According to the terms of international scientific collaboration, this required concurrence of the Government. The subject was, therefore, discussed in an informal meeeting of concerned Indian and U.S. scientists (Roger Revelle, Peter Rogers, and the author) with the then Member, Planning Commission, Dr. M.S. Swaminathan, attended by the then Secretary, Government of India, Water Resources, Shri C.C. Patel. Shri Patel promised to get the field tests carried out. The NCIWRD (1999) noted that they should be carried out urgently.

11. A small structure was constructed on a small monsoon *nallah* (small stream) in the small village of Lava ka Baas in the Alwar district of Rajasthan by the villagers, supported by nongovernmental organizations. The officials planned to dismantle it because they considered it unsafe. The Centre for Science and Environment, New Delhi, headed by an old student of the author, and which is carrying out exemplary social service in this area, invoked the support of the author and a few other scientists supporting the local action (Swaminathan et al. 2002). On examination of the structures, it was shocking to see how unscientific the activity was, understandably because of the ignorance of the villagers, even if the structure was small. It was saved by the author and Swaminathan from the fury of the official agencies, but the structures were washed away in the first rains.

12. World Bank held a Conference of some scientists working on the problem in 1986 in Washington, DC, to which the author was invited to address. The Government is seized of the problem as brought out in GOI (2002, p. 121).

6

Revolutionizing the Development of India's Waters

6.1 Introduction

It was shown in Chapter 5 that a revolution is needed in the concepts, policies, technology, planning, and management of India's waters. We demonstrated the possibility of some novel technologies to allow us to revolutionize the development and management of India's waters. Because development has to be undertaken in terms of river basins, implications and future perspectives in terms of them have to be developed. We first briefly consider the Indus basin, whose development has been undertaken on a pioneering basis. This is followed by the Ganga–Brahmaputra–Meghna (GBM) basin, which is the key to the ushering of the revolution. It has to be considered in terms of the riparian countries because it is an international river basin even though India is the dominant component. Study of the peninsular region follows.

These are merely conceptual ideas. They have to be worked out in terms of policy planning through modern advances in water resources systems and a set of system studies. Some work has been done in this context for the GBM basin and the Indus basin, but it did not consider the proposed revolutionary advances (Chaturvedi 1987). Therefore, these have to be further advanced in terms of the latest developments in Societal Environmental Systems Management (SESM), as discussed below, where an overview of the development of these basins in these proposed terms is provided and the historical and current development is briefly brought out.

6.2 Development of Indus Basin

The Indus basin provides an outstanding example of human environmental interaction shaped by water and its management. The Indus basin is a very arid region, with an average precipitation of about 240 mm. It is almost concentrated in the three monsoon months of July, August, and September. Similarly, there is variation from north to south, with the southern delta region being almost devoid of any rain. Most of the river flows in the summer are from the snowmelt in the Himalayan glaciers. Indus, with a discharge of 20.715 Mha·m, has five major tributaries meeting it from the east. It was an arid barren area inhabited by marauding tribes. Irrigation was practiced from prehistoric times, but it got a boost when Muslim invaders from the even more arid west attained control of the area during the thirteenth century. However, organized development and management started only when the area was brought under the British domain in the mid-nineteenth century. Development of irrigation and settlement of the tribes as agriculturists were the centerpiece of British policy and activity. A large number of inundation canals had been built in earlier times, which had almost silted up. The initial effort was to restore them, but gradually, headworks were constructed on all major rivers (Figure 6.1). An extensive and impressive canal system was developed to divert the water from these major rivers to irrigate the adjoining lands. The pre-1947 irrigation system of the Indus basin covered 10.5 Mha (26 million acres), diverting 90 billion cubic meters (BCM) [75 million acre feet (MAF)] of the 102 BCM (83 MAF) of water, on average, that flowed annually to the Arabian Sea. In addition, there were large numbers of wells, mostly animal-powered shallow percolation ones, which irrigated a further 2 Mha (4–5 million acres) annually. The remarkable fact is that this large irrigation system [the second largest in the world, next to China's and larger than that of the entire United States, the next most irrigated country,

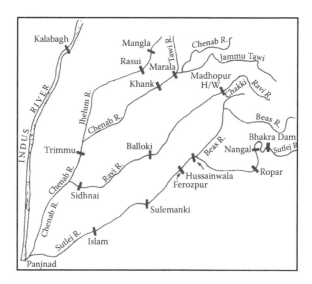

FIGURE 6.1
The headworks built on the various rivers of the Indus basin during the British rule. (From Rangachari, 2006.)

which irrigates 9.3 Mha (23 million acres)] did not have any storage reservoir. Irrigation was entirely dependent on natural river flows, which made it uncertain and limited. There was no flood control or energy generation.

Partition of the Indus basin between India and Pakistan, with the division of British India into Pakistan and India, was a nightmarish event, uprooting millions of people and causing much bloodshed and lasting hostility between the two countries. There were an estimated 12–17 million people uprooted then, with about half to 1 million casualties. Practically, the entire Hindu–Sikh population of West Pakistan migrated to India, and most of the Muslims of East Punjab went over to West Pakistan.

All the important rivers of the Indus system, including the main river, originate in India or pass through India before crossing over to Pakistan. A water dispute between the two parts of Punjab on the use and control of the waters of the Indus system arose overnight as British India was partitioned and two new independent nations were born.

The cultivable land in the basin was 65 million acres, of which 26 million acres remained in India and 39 million acres went to Pakistan. Out of the 28 million acres under irrigation at the time of the Partition, only 5 million acres was left in India. In other words, of the 26 million acres of cultivable land that went to India, only 5 million acres (i.e., less than one-fifth) was irrigated land. This 28 million acres of irrigated land was using some 75 MAF of the Indus system waters, of which, as of 1947, about 66 MAF was used in the Pakistan canals and the remaining 9 MAF in the Indian canals.

The irrigation works were divided, and settlement of the Indus dispute became a most important and urgent issue. This was brought about through the offices of the World Bank. On account of the characteristic features of the Indus system, a neat solution emerged. The basin was cut into two, following the line of partition: Indus and its tributaries in Pakistan, along with their waters, belonged to Pakistan, whereas the three tributaries in India, namely, Sutlej, Beas, and Ravi, along with their waters, belonged to India. A historic Indus Treaty was arrived at between India and Pakistan. It has been studied in detail (Michel 1967).

Under the Treaty, the waters of the eastern rivers, the Sutlej, the Beas, and the Ravi, totaling about 4.069 Mha·m (33 MAF), were fully allocated to India, and those of the western rivers, the Chenab, the Jhelum, and the Indus, totaling about 16.646 Mha·m (135 MAF), were allocated to Pakistan with some limitations.

Reconstruction of the Indus basin in India and Pakistan followed immediately after the settlement. The development and management of Indus in Pakistan became an affair of the Western powers, technically and financially, as Pakistan was their ally.

Development of the waters of India was taken indigenously after Independence, and development of the Indus waters was taken up with utmost priority. It was the leading piece of the new India's water resources development. Major dams on all the three tributaries of Indus in India were constructed with utmost urgency, and a vast network of canals serving the Indus basin in India was developed.[1] With the availability of hydroelectric power, development of the groundwater through small farmer-owned tubewells became the revolutionary activity in the management of water. It gave the farmer control over the water. Agricultural revolution with the new wheat seeds took place. Water resources development and management in India took a new direction, under the inspiration and the pioneering development in the Indus basin in India, both in terms of surface water development through dams and canals and, conjunctively, with groundwater development. However, the colonial canal water management policy continued. It was the development of groundwater, under the control of the farmer, that ushered in the agricultural revolution. The development of the water resources of the Indus basin in India, with the three major dams and canal network, is shown in Figure 6.2.

Two pioneering developments in the field of water resources development and management followed. With the development of extensive irrigation in the arid Indus basin, a major problem of salinity had been emerging even in British India. Attempts to solve it had been made by the Indian scientists, but not much success was achieved. With Pakistan joining the Central Treaty Organization, the salinity problem management in Pakistan was undertaken by US scientists lead by Roger Revelle (Report on Land and Water in the Indus Basin, 1964).

Second was systems planning of water resources management. The science was still emerging. A pioneering systems study of the Indus basin in Pakistan, although simplistic, was undertaken (Lieftinck et al. 1968). This encouraged some U.S. scientists to undertake a systems planning study of the developments of the Indus basin in India, in collaboration with some Indian economists. Again, although simplistic, it motivated the policy-making economists in the India Planning Commission to authorize this author and his colleagues to undertake systems planning for the Indian Indus basin. Although elementary, it became this author's starting point of systems planning in India (Chaturvedi 1973).

The Harvard group became the consultants for scientific development of the water resources of East Pakistan, later Bangladesh. The Ford Foundation in India provided developmental support to the scientific capability of water resources systems planning in India, encouraging this author to embark in this newly emerging field. We undertook collaborative studies with Harvard scientists to establish indigenous scientific capability in India (Chaturvedi and Rogers 1985). However, real-life studies of the Indus basin eluded the author, as the Nehruvian culture of promotion of science was totally replaced by a culture of making money. This was brought to the author's experience dramatically by the Punjab engineer deputed for training and education in the context of Punjab water's systems studies as discussed in Chapter 8.

Third, the policies developed for the GBM basin apply to the Indus basin, as the hydrologic and physiographic conditions are similar in principle, as

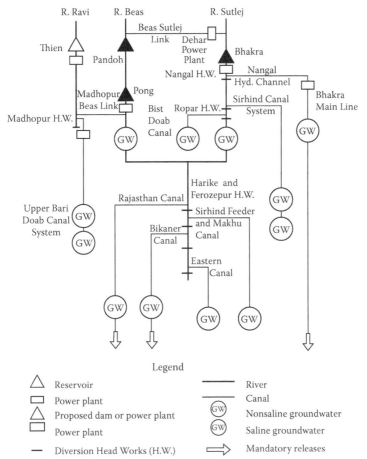

FIGURE 6.2
Schematic diagram of the Punjab water resources system. (From Chaturvedi, M.C., *Water Resources Systems Planning and Management*, Tata McGraw-Hill, New Delhi, 1987.)

discussed above. However, they are vastly different in detail. The Himalayan energy potential can be developed by storing the energy in the Central highlands through the Chaturvedi Water Power Machine. The waters can be stored in vast alluvial plains. It is deemed that considerable advancement can be made through novel technologies for the Indus basin as well, duly adapted to the Indus basin conditions.

Fourth, the climate change specter looms all over the country, but it poses a particularly serious challenge for the Indus basin in view of its large water supplies from the Himalayan glaciers, estimated to be about 180 BCM of water each year. Even with corrective action taken at a global level, glacial retreat will continue for at least half a century. River flows will increase, raising the likelihood of flash floods and exacerbating already acute drainage problems. In the second half of the twenty-first century, there is the likelihood of a dramatic decrease in the river flows, conceivably by more than 30%. This major permanent reduction in runoff will have major consequences on the economy and the life of the people. Attempts to deal with this severe environmental impact should be started urgently and in earnest by studying the subject in terms of the new concepts and technologies, as proposed by the author.

6.3 Development of GBM Basin

Nehru declared, "Ganga above all, is the river of India which has held India's heart captive and drawn uncounted millions to her banks since the dawn of history. The story of the Ganga from her source to the sea, from old times to new, is the story of India's civilization and culture" (Nehru 1947).

Indeed, revolutionizing the development and management of the GBM is the heart of the proposed revolution of India's waters. It has been brought out in detail in an accompanying study (Chaturvedi 2011d). It is presented here briefly, just to emphasize and bring out the proposed revolution in the development and management of India's waters, as brought out in principle in Chapter 5.

The layout of the GBM basin is shown in Figure 6.3 (Figure 5.3 reproduced). The GBM river system constitutes the second largest hydrologic region in the world. The total drainage area of the GBM basin is about 1.75 million km^2, stretching across five countries: Bangladesh, Bhutan, China, India (16 states in the north, east and northeast—in part and fully), and Nepal. Whereas Bangladesh and India share all the three rivers, China only shares the Brahmaputra, Nepal only the Ganga, and Bhutan only the Brahmaputra. Growing at an average rate of about 2% per annum over the past decade, the estimated population of GBM had reached a level of about 600 million by 1999. India's share in the population and the area of GBM region are 76% and 63%, whereas the corresponding shares of Bangladesh are 21% and 7%, respectively. Nepal, whose almost entire territory is in the Ganga system, has an 8% share of the GBM area and 3.5% of the GBM population. About 10% of the world's humanity lives in this region, which contains only 1.2% of the world's land mass. The population is estimated to stabilize at about a billion.

The GBM region is characterized by endemic poverty—being home to about 40% of the total number of poor people residing in the developing world. The performance of the region with respect to such social indicators as economic growth, education, and health is disappointing (Table 6.1).

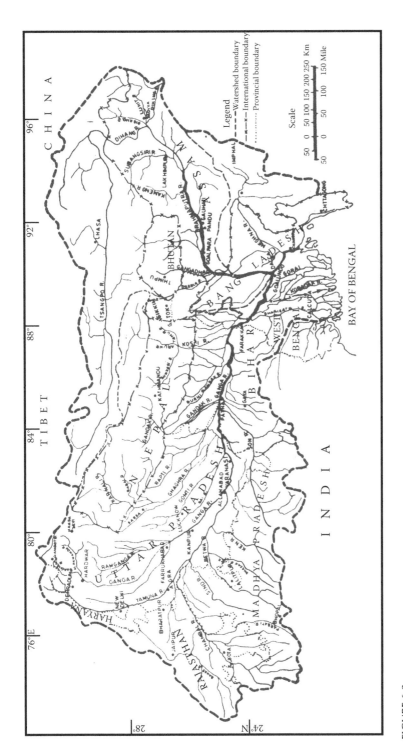

FIGURE 6.3
The layout of the GBM basin.

TABLE 6.1

GBM Region—Socioeconomic Indicators

	Bangladesh	India	Nepal
Population (million) 1998	128	987	24
Annual population growth rate: 1995–2000 (%)	1.9	1.8	2.5
Access to safe water (percentage of population) 1995	84	85	59
Access to sanitation (percentage of population) 1995	35	29	20
Adult literacy rate (percentage of people 15 and above) 1997	50(M)	67(M)	56(M)
	27(F)	39(F)	21(F)
Female (as percentage of labor force) 1998	42	32	40
Arable land (hectares per capita) 1994–1996	0.07	0.17	0.13
Per capita commercial energy use; annual (kg OE)	197	476	320
Per capita electricity consumption (kWh) 1996	97	347	39
Population below national poverty line (%) 1990s	48	37	43
Per capita GNP (US$) 1998	350	430	210

Source: Ahmad, Q.K., Ahsan, U.A., Khan, H.R., and Rasheed, K.B.S., in *Ganges–Brahmaputra–Meghna Region—A Framework for Sustainable Development*, edited by Q.K. Ahmad, A.K. Biswas, R. Rangachari, and M.M. Sainju (eds.), The University Press Ltd., Dhaka, 2001.

Note: kg OE = kilogram of oil equivalent; kWh = kilowatt-hours.

Despite the poor socioeconomic status of the region, it has rich natural endowment of land, water, and energy. Nearly 45% of the land is arable, but per capita availability of land is very small—around one-tenth of a hectare. Owing to population increase, per capita availability has been going down continuously.

The GBM region is a very rich water region. It is widely recognized that water is the most important vector of development. The average annual water flow in the GBM region is estimated to be around 1350 BCM, nearly half of which is discharged by the Brahmaputra. The three rivers constitute an interconnected system—discharging into the Bay of Bengal. Compared to the world's average annual availability of 269,000 m^3/km^2, the availability in the GBM region is 771,400 m^3/km^2, nearly three times the world's average. In addition to surface water, the GBM region has an annual replenishable groundwater resource of about 230 BCM. The land-water characteristics, however, vary widely over the region. The abundance of water in the GBM region is the principal driver of development as well as of woes for millions of people in the region. The shared river system can be optimally developed only through collaborative effort among the concerned nations.

An important characteristic, with implications for creative development and current impacts, as developed in Chapter 5, is the highly concentrated flow of water, as shown in Figure 5.1. About 80% of the annual rainfall occurs during the 4–5 monsoon months, often concentrated in heavy spells over 10–12 days. The flow in August may be as high as 15 times the low flow that occurs in February. It is further compounded by the fact that it increases as one moves downstream. Flooding, which increases in magnitude as one moves down, is a formidable challenge. The GBM rivers convey an enormous amount of sediment load from the mountains to the plains, which compound the adverse effects of floods. The temporal concentration also poses a central and difficult challenge for development of the water resources. A central conclusion is that while we will formulate some general and novel guidelines for development, which will enormously increase the current potential for development, detailed planning has to be undertaken urgently and will be nation centric. Some action in this context has been initiated, as discussed later, which will provide an important guideline.

6.4 Revolutionizing Developmental Strategy of GBM Basin

The GBM is divided internationally, almost, along the major physiographic divisions. The Himalayan region is essentially in Nepal and the delta region is dominantly in Bangladesh. The alluvial plains are almost entirely in India, and the central highlands are entirely in India. The proposed novel technologies contribute to the development of collaboration, as they provide highly enhanced benefits to the entire region and each country, individually and collectively.

The entire GBM faces a most challenging socioeconomic developmental–environmental sustainability problem. Our perspective is focused on the GBM being one socioenvironmental entity that is collectively fighting to establish its due place in the human community. It has a long cultural history. It provides very complimentary opportunities if environmental management is undertaken without bedeviling it with spatial political divisions. It is not an idealistic imposition but follows from the simple basic fact that a river basin is one environmental entity and does not know political administrative boundaries. Second, from the very perspective of development of the waters, it is seen that the dominant activity is in one political unit, leading to easy international collaboration and implementation. Having visualized development in this context, development has to be approached under national terms.

The characteristic division of the GBM basin along the four distinct physiographic regions indicates their consideration along them, which is also synonymous with collaborative Nepal–India and India–Bangladesh development, leading to easy implementation. Overall, coordinated development could be considered later to further optimize these specifically indicated developments.

6.5 Development of Himalayan Region

The Himalayan region poses serious challenges to development. However, the example of Switzerland, which once was the poorest region of Europe but is now the world's richest country, provides inspiration.[2] Nepal is the dominant Himalayan region, but the policy is the same for the entire region. As far as the development of the region is concerned, we follow the policy proposed by Nepal, in principle (Thapa and Pradhan 1995; Malla et al. 2001). We add to it the proposed novel policies of developing the major rivers, starting with the Indian part. Hopefully, Nepal will be attracted by them and will follow.

6.5.1 Geographical and Ecological Setting

The Himalayan region has five distinct regions from south to north—Terai, Siwalik (or Churia), Mahabharata, Inner Terai (or Bhitri Madesh), and the Himalayas, as particularly brought out by focusing on Nepal (Figure 6.4). The southern flank of the Himalayas drains into the Ganga or Ganga River system, whereas the northern side forms part of the Tsangpo/Brahmaputra. There is another distinctive ecological setting along the major rivers, the tributaries of Ganga, which are, from west to east, Yamuna, Ganga, Ramganga, Sarda or Mahakali, Ghagra or Karnali, Gandak or Gandaki, and Kosi.

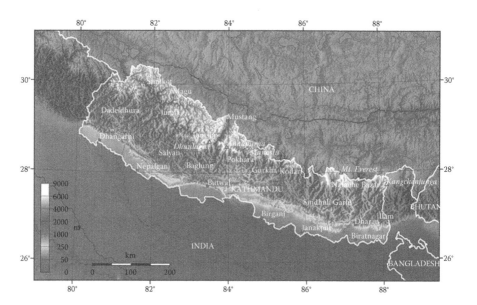

FIGURE 6.4
Nepal.

6.5.2 Socioeconomic Setting

Poverty is endemic and widespread in the Himalayas. Forest resources are under severe strain. This has led to serious ecological disturbances. Traditional agriculture, nascent industrialization, unemployment, and underdevelopment are the chronic problems of the Himalayan economy. There is also widespread inequity in distribution of income and wealth.

6.5.3 Resource Base

Compared to the world's 269,000 m^3/km^2, Nepal's annual average water availability is 1,182,000 m^3, which is of the same order as that in the Indian Himalayan region. This is about four times the world's average. The Nepal rivers contribute about 41% of the total runoff and 71% of the lean flow of the Ganga. Nepal has a dense network of more than 6000 rivers and rivulets, with a total runoff of about 224 BCM. The groundwater potential in the Terai, in Nepal, is estimated to be about 12 BCM.

The Himalayan rivers may be further grouped into three categories, depending on their origin and discharge: Himalayan rivers, Mahabharata rivers, and Siwalik/Churia rivers. The Himalayan rivers are perennially snow-fed rivers. Although the flow in the dry months is only 5%–10% of that in the wet months, it is still a significant amount for such a period. Yamuna, Ganga, Ramganga, Sarda/Mahakali, Ghagra/Kosi, Gandak, and Karnali fall into this category.

The Mahabharata rivers originate from the Mahabharata range, and the dry season flow is because of the dry season recession. Some of the main rivers in this category are Kanaka, Kamala, Bagmati, Rapti, and Babai. Eighty-seven percent of the flow occurs during the wet months, between June and October. The basin areas of these rivers, especially the river valleys, are densely populated, and therefore, feasible storage is not possible. The Churia rivers, originating from Siwaliks, are seasonal rivers and are usually dry during the lean season. They are used to flash floods.

6.5.4 Temporal Variation

The distribution of flows is very uneven in space and time. In the case of both snow-fed and non-snow-fed rivers, 24%–27% of the annual flow occurs in the month of August, and only about 1%–2% occurs in February. The flow in August is as high as 15 times the low flows in February.

6.5.5 Water Resources Development Approach

A three-pronged approach has been proposed by Nepalese scientists: (1) decentralized programs and technologies to meet local needs; (2) medium-sized projects to meet the national needs of energy, food, and water supply; and

(3) projects of a multipurpose nature whose benefits extend beyond national boundaries (Malla et al. 2001). The same applies for the Indian Himalayas.

6.5.6 Development Policy

Life is hard in the Himalayas. Local development is the first issue. Environmental management is embedded in the socioeconomic development, which is the first priority. In this context, drinking water, which is difficult to obtain, is the first priority. The usual arrangement is to obtain it from some hill stream or hill spring. Attempts at irrigation have also been made by bringing water from river streams along the hill contour over long distances for drinking purposes and irrigation. As one reviews them, for all their simplicity, one marvels over the tremendous effort mankind makes for obtaining water.[3]

In the first instance, we will have to improve them; ultimately, we will be able to provide water for drinking purposes and irrigation in abundance as the storage projects are developed. This also emphasizes the urgency of the development of the storages of all sizes, as much as economically feasible. However, much preparatory work, such as construction of roads and watershed management, has to proceed. Thus, an urgent need for organized long-term planning, in sufficient detail, and immediate action is indicated.

The Himalayan region is very rich in hydroelectric potential but, at the same time, has a very fragile ecosystem. The Himalayan region faces one of the most challenging opportunities for economic and environmental development. Close collaboration between Nepal and India, as well as among all the three countries of the region, is a central issue.

6.5.7 Revolutionizing the Hydrolectric Development

The hydroelectric development has been undertaken essentially in the Indian Himalayas so far. It was demonstrated in Chapter 5 that it should be revolutionized, both for the run-of-river and the storage projects, considered in terms of an integrated water and power system over the entire river basin, and even in the subcontinent in terms of the Chaturvedi Water Power Machine.

A hydroelectric potential of 128,700 MW at 60% load factor has been estimated in the Himalayas, of which 83,000 MW is in Nepal and 45,700 MW in India. Of the total, 50% is estimated to be through run-of-river schemes and the rest through storage projects. Both are currently designed based on 90% water availability. The Indian potential is dominantly in the Brahmaputra basin. As for that of Nepal, it is estimated that only 40,000 MW is economically feasible (Malla et al. 2001). Only about 1.5% has been developed so far.

As we have argued, the concepts of development and these estimates are vastly incorrect. Development has to be revolutionized according to the following principles and approach. First, the low flows should be conserved based on environmental considerations and should not be diverted to generate hydropower, as what is being done currently. Instead, the energy of the high flows should be fully developed up to the economic levels, according to the Chaturvedi Water Power Machine. The potential and economic benefits, according to the proposed revolution, is much higher by several orders of magnitude. The immense hydroelectric potential of the GBM must be developed with utmost urgency because it will not only provide much needed power but will also provide employment opportunities, particularly to the people in the Himalayas. The Chaturvedi Water Power Machine provides phenomenal increase, not only in the overall potential but also in the value of water power for development. It may be brought out that the storage projects can only utilize a fraction of the total water resources, estimated to be about 38% (Thapa and Pradhan 1995, p. 135). Thus, the Chaturvedi Water Power Machine, which can also vastly contribute to the use of the potential of the unstorable waters, will further enhance the benefits. It depends vastly on a collaborative India–Nepal development of the GBM waters. As the scheme goes, Nepal will undertake the development with utmost enthusiasm if the developmental revolution is brought to their attention.

Besides the psychological barriers, cost sharing has been a point of dispute. It will be seen that with the Chaturvedi Water Power Machine, the conventional perspective of cost sharing is completely lost and a new picture emerges. Even more importantly, cost sharing may be worked out on the principle of contributing to the development of the region and not necessarily on the basis of opportunty costs alone, particularly in terms of the proposed enhancement of the basic principle of development of water, power, and environment conservation and the basic principle of eradication of the abysmal backwardness of the region, which has a long history of cultural integration. Committed creative integrated development of the region, irrespective of the politcal boundaries and divisions, is the central issue, and it is in the interest of both the countries.

It must be emphasized that the development of these rivers is a very major and complex engineering activity. It will also require very huge financial and engineering inputs. India, much less Nepal, does not have the potential to undertake the development in terms of the modern, much less the revolutionary, proposals. The first step, therefore, is undertaking collaborative development of one major river system, say, Mahakali/Sarda, involving the scientific professional–academic expertise of both countries. It is time that a new relationship, renewing the historic close relationship of the Indian subcontinent, expressed by nature and our sages, is re-established through the Chaturvedi Water Power Machine.

It may be emphasized that soil conservation is the first and most important issue. Under the geophysical and the hydrologic conditions, the Himalayan region suffers from one of the world's most serious soil erosion. Soil conservation is valuable in its own right. It is, further, most valuable in water resources development, as it contributes to conserving the valuable storage potential created at huge costs.[4]

6.6 Development of Alluvial Plains

The alluvial plains are the heart of development of the GBM basin and India. They are essentially in India. The development of the region can be entirely transformed through the proposed revolution in water resources development and management strategies through the Chaturvedi Ganga Water Machine. The basic revolutionary concept is that the land and water are one system, even if water is variable over time and space. We use the land to store water in times of high input and retrieve it during times of need, conjunctively with the temporal spatial variability, locally and regionally.

We emphasize meeting the environmental and domestic requirements as the first priority. The water demand is trivial, in contrast to the agricultural sector demands. But it emphasizes the fact that the lean period flows, after the monsoons, cannot be diverted, as what is being done currently. With the Chaturvedi Ganga Water Machine, we ensure abundant water and that too in the hands of the farmer, around the year. We transform the Kharif and Rabi agriculture. We also ensure abundant flows all year round in the rivers, ensuring the best environmental state.

Management of floods, with construction of embankments, has posed serious problems of local hardships, particularly as the flow intensity increases as one moves east. Shifting to Chaturvedi Ganga Water Machine technological policy, with emphasis on the use of groundwater in the alluvial plains, may minimize the problems. However, the appropriate developmental policy demands much creative action, instead of being hooked to canal irrigation. It must be emphasized that the Indo-Ganga basin has vastly varying characteristics as one moves from west to east. The technology of the Indus basin is not suitable for the eastern Ganga basin, the Kosi region, or further east to the delta region.

It must be emphasized that the activity requires scientific regional planning and management, involving close collaboration of the local community. Enthusiastic involvement of the local people is important, but enthusiasm is no substitute for science. We are not talking of "local water resources development and management" as brought out by NCIWRDP 1999 but of regional development. The sad experience of Lava ka Baas discussed in Chapter 5 underscores this issue.

6.7 Development of Vindhyan Region

The story continues in the Vindhyan region. Through the Chaturvedi Water Power Machine, abundant energy is developed. The Chaturvedi Water Power Machine ensures that the land is provided with adequate water through development of the land–water alluvial system and through small-capacity storages, which provide water under the control of the farmer. Valuable interbasin transfer to Penisular India also takes place.

6.8 Development of Brahmaputra–Meghna Region

The Brahmaputra region has rich water and hydroelectric potential. The Chaturvedi Water Power Machine offers valuable enhancement of the energy potential, also contributing to the enhancement of the water potential of the entire India, through interbasin transfer to Peninsular India, as discussed for the Ganga basin. Similarly, the Chaturvedi Ganga Water Machine ensures revolutionary transformation of the agricultural sector. Detailed investigations are required for actual implementation, particularly as the land–water system varies widely over the GBM basin.

6.9 Development of Delta Region

6.9.1 Introduction

The delta is dominantly in Bangladesh. We follow the development of Bangladesh as proposed by some of their scientists (Ahmad et al. 2001). This applies to the entire delta with local variations all over.

Bangladesh is a small, poor country with a large population. The total area of the country is 147,570 km², 6.7% of which consists of rivers and inland water bodies. Some 88% of the country's total area belongs to the GBM region. The estimated population of the country as of 1999 is about 128 million with a population density of about 860 persons/km². The country's population is estimated to exceed 176 million by 2025, when the population density will rise to about 1200 persons/km². Currently, the urban population accounts for about 20% of the national population, which is estimated to rise to about 53% by 2025.

The water ecosystem of Bangladesh comprises the tributaries and distributaries of three major river systems: the Ganga, the Brahmaputra, and the Meghna, and numerous perennial and seasonal wetlands (Figure 6.5). The following particular hydrological features may be noted: (1) Of the total annual stream flows of Bangladesh, 85% occurring in between June and October, 67% is contributed by Brahmaputra, 18% by the Ganga, and 15% by Meghna, and other minor rivers. About 93% originates outside the country. (2) The annual volume of surface water in Bangladesh would form a lake of

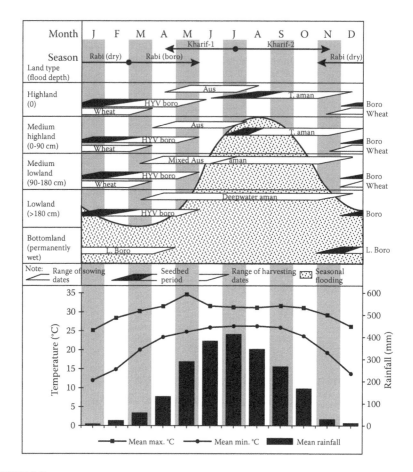

FIGURE 6.5
Map of Bangladesh highlighting the river systems. (From Ahmad, Q.K., Ahsan, U.A., Khan, H.R., and Rasheed, K.B.S., in *Ganges–Brahmaputra–Meghna Region—A Framework for Sustainable Development*, edited by Q.K. Ahmad, A.K. Biswas, R. Rangachari, and M.M. Sainju (eds.), The University Press Ltd., Dhaka, 2001.)

the size of the country having a depth of 10.2 m. (3) Bangladesh has to drain water from an area which is 12 times its size. (4) One-third of the area of Bangladesh is influenced by the tides in the Bay of Bengal.

Bangladesh has three broad types of land: floodplains (80%), terraces (8%), and hills (12%). The precipitation is dominated by monsoon characteristics. The average annual rainfall varies from 1200 mm in the extreme west to over 5000 mm in the northeast, with 80% of the monsoon being concentrated in the four monsoon months: June–October.

Bangladeh is predominantly an agrarian society. The country has about 8.74 Mha of cultivated land, which is about two-thirds of the total area. Of the net cultivable area, 33.3% is single cropped, 45.0% is double cropped,

11.5% is triple cropped, and 10.2% is cultivable waste and presently fallow. The overall cropping intensity is 176%. The three cropping seasons approximately concide with the three meterological seasons: Kharif I (pre-monsoon), Kharif II (monsoon), and Rabi (winter or dry). Aus, Aman, and Boro are the three rice varieties grown, respectively, in these three cropping seasons. The crop calender (grain based) is presented in comparison to the temporal distribution of rainfall and temperature in Figure 6.6. Aman is the leading rice crop, occupying about 56% of the total area under rice, followed by Boro (27%), and Aus (17%). A notable aspect of the pattern of growth in crop agriculture during the past two decades has been the increasing area covered by the dry season high-yielding variety Boro rice—a trend that is likely to continue.

6.9.2 Problems of Water Resources Development

The critical physical/natural problems that Bangladesh faces are flooding, riverbank erosion, sedimentation/siltation strategy, water scarcity, and salinity management.

Floods are a recurring phenomenon in Bangladesh. Even in a normal year, 20%–30% of the country is flooded. About 60% of the country is submerged by a flood of about a 100-year return period. Up to 80% of the country is considered flood prone.

Most of the rivers of Bangladesh flow through unconsolidated sediments of the GBM floodplain and delta. The river banks are susceptible to erosion. The GBM rivers flow through well-defined meander belts on extensive floodplains where erosion is heavy. River–land interaction is severe and varied.

Bangladesh is the outlet of all major upstream rivers and the average annual sediment load that passes through the country to the Bay of Bengal ranges between 0.5 and 1.8 billion tons. All rivers in Bangladesh are alluvial and highly unstable. There is very heavy siltation leading to increasing vulnerability to floodplains.

Salinity is a major problem in Bangladesh. Almost all the rivers of Bangladesh combine to form a single broad and complex estuary. The greatly diminshed flow in the dry season allows salinity to penetrate far inland through this esturine river system. The coastal zone directly affected by salinity is extensive and is inhabited by a large population.

Studies of water availability and demand are complex. According to the latest estimates, water demand for all purposes (domestic, irrigation, fisheries, navigation, industrial, and salinity control) for the year 2018 was estimated at 24,370 million cubic meters (MCM) during the critical dry month of March. The supply from regional and domestic sources in terms of both surface and groundwater was estimated at 23,490 MCM, producing a short fall of about 880 MCM or about 3.5% of the estimated demand. It is estimated that the difference will decrease with estimated improvement in the efficiency of use. Thus, though the supply–demand picture is not of too much

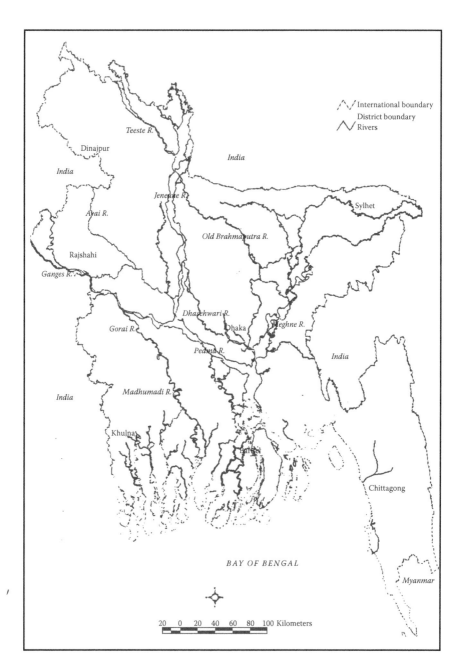

FIGURE 6.6
Crop calendar in relation to seasonal flooding, rainfall, and temperature. (From Ahmad, Q.K., Ahsan, U.A., Khan, H.R., and Rasheed, K.B.S., in *Ganges–Brahmaputra–Meghna Region—A Framework for Sustainable Development*, edited by Q.K. Ahmad, A.K. Biswas, R. Rangachari, and M.M. Sainju (eds.), The University Press Ltd., Dhaka, 2001.)

concern, spatial imbalance will, however, not match the overall regional balance. This can, however, be overcome through judicious planning because the imbalance is minor.

Bangladesh presents difficult challenges, but with the rich land–water resources, they can be met satisfactorily.

6.10 International Interaction in GBM Basin
Current Scene—Nepal and India

The position regarding the Nepal–India interaction is brought out as stated by some Nepalese scientists involved in some attempts at interaction. Writing about riparian cooperation, experience and issues, it was stated that "In a span of five decades, Nepal has entered into several agreements with its lower riparian, namely India, in harnessing the Himalayan rivers (the Kosi, Gandaki, and Mahakali) that covered areas like irrigation, flood control, and hydropower generation. The outcome of a series of project agreements between the two countries has not been satisfactory, vitiating bilateral relations from time to time. Uncooperative stances like irregular exchange of information, differing opinions on the project, and mutual distrust and suspicion lead to slow progress on creating an enabling environment for joint water resources development." (Malla et al. 2001, p. 156)

It is time that a climate of close collaboration is established. Nepal has difficult sociopolitical problems and also has grievances toward India, yet initiative and perseverance has to be displayed by India, as management of Nepal–India waters affects vast humanity in the three riparian countries.

We believe that the Chaturvedi Water Power Machine can open up a new future. According to it, the immense water potential of Nepal's waters, in terms of energy and water, can be developed, increasing the benefits to Nepal immensely. As Nepal appreciates its potential, the initiative will be taken by Nepal. The entirely new perspective of India is to facilitate this change in Nepal's perspective, offering to develop collaboration, as Nepal deems appropriate. The entire issue of sharing of benefits becomes secondary.

6.11 International Interaction in GBM Basin
Current Scene—India and Bangladesh

Conflict with Bangladesh (formerly East Pakistan) started right from the achievement of Independence. The dispute principally is about sharing the

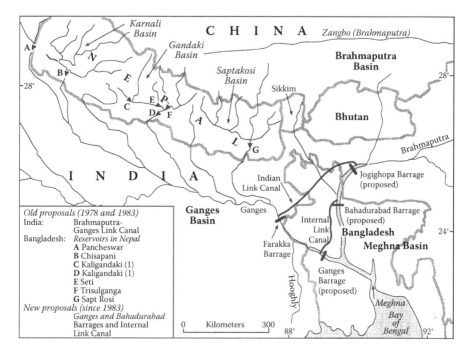

FIGURE 6.7
GBM conflict resolution proposals. (From Crow, B., *Sharing the Ganges*, Sage Publications, New Delhi, 1995.)

low flows of Ganga. It was agreed that both countries develop proposals to augment the low flows. India has proposed diverting water from Brahmaputra to augment the low flows upstream of Farakka Barrage. Bangladesh has proposed that water transferred by India from Ganga should be compensated from the Ganga basin itself through additional capacity that will be provided in the storages to be developed on the high dams in the Himalayan rivers to make up for the transferred Ganga water to meet the additional water demands of Bangladesh. The two proposals are shown in Figure 6.7. Each side has rejected the other side's proposal. Meetings continue as "dialogues of the deaf," as Mehta has called them.[5]

6.12 GBM Conflict Resolution

An effort was made by Mehta and the author, while both were working at the University of Texas at Austin, USA, in 1989–1990, to undertake collaborative scientific development of the GBM basin by scientists working to work out a policy of development according to the modern advances of systems planning. Considerable work focusing on India had been carried out by the author, and this encouraged Mehta, who had been leading the India side for some time in these discussions. A workshop was organized in Australia as part of a scientific gathering to bring out the modern scientific advances in this field and the necessity of undertaking the development according to these modern advances (Eaton 1992). However, no progress could be made. Some well-meaning persons from the three countries continued interaction. These have been brought out in a number of books (Verghese and Iyer 1992; Thapa and Pradhan 1995; Ahmad et al. 1994, 2001a,b). Although valuable, as described in the foreword by Panandikar (1992), Director of the Center for Policy Research in New Delhi, "These studies are addressed to the lay reader and are not intended to be technical treatises."

Although the studies undertaken so far provide valuable insight, the subject has to be studied in terms of the modern advances backed by scientific participatory interactive modeling. Lack of data is often quoted as a problem. For policymaking, much can be accomplished based on the available data through modeling. We therefore propose a different approach. We believe that scientific collaborative study is undertaken by the scientists and professionals of the three countries. Because this would not be possible right away, a three-track approach is proposed. First, we work out the development of the GBM according to these novel concepts and technologies for the region in our control. For example, the Himalayan development of Yamuna–Ganga–Ramganga can be undertaken based on the Chaturvedi Water Power Machine. The development of the alluvial plains, the Vindhyan region, and the Brahmaputra–Meghna region can be planned out on the basis of these revolutionary advances, and implementation should be undertaken by India, as it is entirely in India's jurisdiction. Similarly, the development of Brahmaputra, in India, particularly the immense hydropotential development, should also be undertaken immediately, as this too is India's domain. We should start implementation accordingly.

Second, we work out the likely development of the entire region in the proposed revolutionary terms and share it with Nepal and Bangladesh. As far as Nepal is concerned, they will hopefully realize the immense benefit they can get through the Chaturvedi Water Power Machine and will themselves take the initiative in implementing their water resources development in the proposed revolutionary terms. It requires very large investments and preparatory work, and the development has to be undertaken one river basin at a time. It requires very close collaborative development by India and Nepal. It will be in Nepal's interest that they join the proposed scientific development on terms best suited to them, but hopefully as equal scientific partners.

Regarding Bangladesh, India should extend all support to Bangladesh as they face very serious problems. The obvious scientific approach is interlinking of Brahmaputra and Ganga, but the decision has to be taken by the Bangladeshi scientists and the people. The subject of sharing of Ganga waters, however, is not a scientific problem, but a political problem. An internal suggestion was made by some Bangladeshi scientists that in Bangladesh's interest, the best approach will be to divert water from Brahmaputra, which has abundant water, at the uppermost point in Bangladesh itself as it enters Bangladesh, to meet the Bangladesh shortfall (Figure 6.7) (Crow 1995).[6] However, it was immediately shot down because it would weaken Bangladesh's legal claim on Ganga waters. It appears that the current meaningless political dialogue will continue until all realize its futility. As has often been said, mankind is not a rational animal.

Bangladesh suffers a very serious problem of arsenic in its groundwater, management of which has to be given top priority. One way of managing it could be by providing a high potential at the uppermost plane by diverting the Brahmaputra waters to Bhagirathi and conjunctively developing the huge potential of artesian groundwater in the GBM basin. The Brahmaputra–Ganga interconnection will also provide excellent navigational interconnection to the entire region. However, the decisions are to be taken by Bangladeshi scientists and the people.

6.13 Scientific Development of GBM

An interesting study of Ganga in the context of global and regional changes in the biosphere over the last 300 years was undertaken by some scientists (Schwarz et al. 1990).

Figure 6.8 represents a time line of Ganga River's development. Attempts to understand the environmental impacts were made. Figure 6.9 presents the 1980 annual hydrograph and, next to it, hydrographs estimated from the change in irrigated area for 1850 and 1900. The enormous effects of human intervention, especially in the low-flow season, are clearly shown. Summary of the conclusions is reproduced below.

"From this cursory review of the historical and current situation, we can conclude that despite continual settlement for almost 4,000 years, humans have not yet caused irreparable damage to one of the world's most valued ecosystems. To be sure, the ecosystem is radically different from what it would have been if humans had not settled in this area. We can also conclude that the most serious impacts are likely to be those in the near future (within 30 years), when the inhabitants will enjoy much higher material standards of living, which will stress the absorptive capacity of the environment.

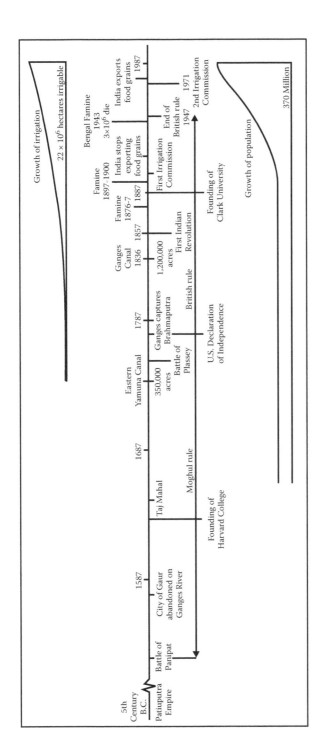

FIGURE 6.8
Time line for the development of Ganga River. (From Schwarz, H.E., Emel, J., Dickens, W.J., Rogers, P., and Thompson, J., in *The Earth as Transformed by Human Action*, edited by B.L. Turner II, W.C. Clark, R.W. Kates, J.F. Richards, J.T. Mathews, and W.B. Meyer, Cambridge University Press, Cambridge, 1990.)

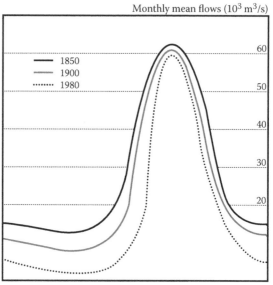

FIGURE 6.9
Mean monthly hydrograph of the Ganga River (hydrograph for 1980 and reconstructed hydrographs for the years 1850 and 1900). (From Schwarz, H.E., Emel, J., Dickens, W.J., Rogers, P., and Thompson, J., in *The Earth as Transformed by Human Action*, edited by B.L. Turner II, W.C. Clark, R.W. Kates, J.F. Richards, J.T. Mathews, and W.B. Meyer, Cambridge University Press, Cambridge, 1990.)

Thus, the Ganga is—and has been for more than 300 years—the second stage of human intervention."

An important conclusion that emerges from the study is the validation of the recommendations made to undertake water resources development management in terms of the SESM (Chaturvedi 2011a). Some efforts were undertaken to carry out a study for the GBM basin, with particular emphasis on Uttar Pradesh, one of the states in the GBM basin, as shown in Figure 6.10, focusing on the development of water resources of the GBM, including the isssues of management of water at field level, and agricultural system as discussed earlier (Chaturvedi and Rogers 1985). Several doctoral dissertations were undertaken (Arya 1980; Asthana 1984; Bhatia 1984; Chaube 1983; Duggal 1975; Gupta 1984; Khepar 1980; Prasad 1981; Singh 1980).[7] It was further extended to include energy issues (Sarma 1985; Thangraj 1987). Further recent advances in terms of intertemporal–interspatial development in the context of environmental systems development have also been undertaken (Kothari 2000; Hasan 2005). However, it should be considerd only as a first exercise of the complex water and energy systems planning. Regional water resources systems development, basically, involves conjunctive consideration

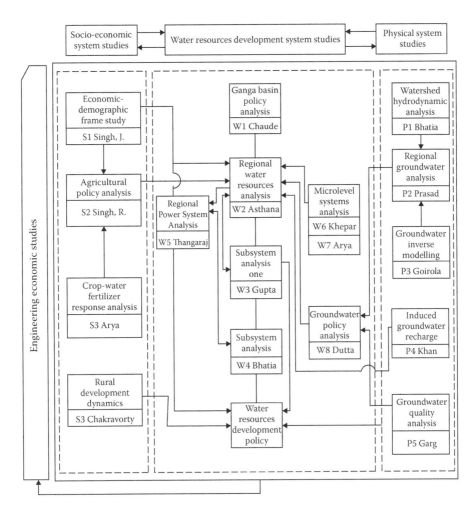

FIGURE 6.10
Scheme of river basin system studies. (The name in the bottom is of the doctoral scholar.) (From Chaturvedi, M.C., *Water Resources Systems Planning and Management*, Tata McGraw-Hill, New Delhi, 1987.)

of socioeconomic issues, physical system issues, and water resources systems development issues in the perspective of SESM as brought out in Figure 6.11 (Figure 5.10 reproduced). Efforts to undertake a study in this context are being made.[8]

Management of river basins is complicated by political issues because most of the rivers are international. The sharing of the GBM basin by Nepal, India, and Bangladesh is a classical example. The three countries are engaged in developing collaborative action, but the scene so far has not been meaningful, or even promising.

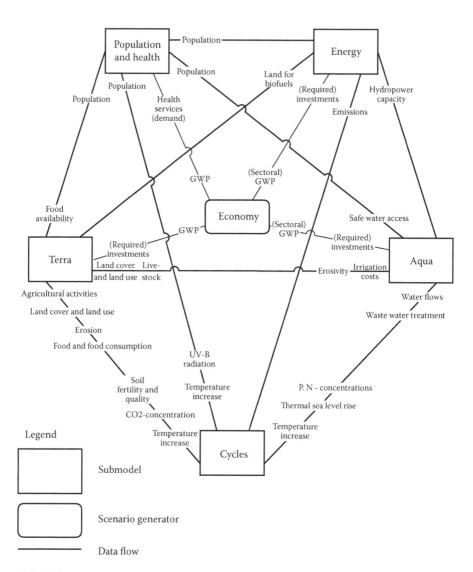

FIGURE 6.11
Societal–environmental systems management. (From Chaturvedi, M.C., *Environmental Systems Management for Sustainable Development*, in press, 2011.)

6.14 Development of Peninsular Region

The development of the peninsular region is like the development of any typical river basin. The role of the groundwater is much less than in the alluvial plains and dependence has to be more on storages, for which potential exists all over the basin. Considerable development has taken place.

The study of the GBM basin demonstrated that the excess waters available in the GBM basin, through the Chaturvedi Water Power Machine, should be transferred to the peninsular region. The magnitude of the waters is a matter of detailed study but is going to be large. The total surplus surface waters of the GBM basin, as per the author's study, have been estimated to be about 275 km^3 (Table 5.1). An important figure to be noted is that it is larger by an order of magnitude than the combined surface flows of Krishna, Pennar, and Cauvery, which amounts to only 97.49 km^3. Even after adding the surface flows of Godavari, which amount to 110.54 km^3, the total amounts to 208.03 km^3. The contribution from GBM is going to be immensely important, particularly when it is almost free in view of the benefits from the pumped storage component. The conclusion is obvious—instead of current interlinking proposals, a study of the subject of development of India's waters should be undertaken on lines proposed in the forgoing analysis, in detail.

Another important point to be emphasized is that although we do not have the possibility of much groundwater component as in the GBM basin, which provides valuable power to the cultivator, besides availability of water, we can still develop the surface waters of the peninsular region from the principle of providing the control in the hands of the cultivators. This can be done by providing small storages in the distribution system, which are in the control of the cultivator unions. The conclusion is that creativity is central to peninsular development as elsewhere.

6.15 Climate Change

The entire Indian subcontinent is considered to face serious problems on account of the climate change. Bangladesh will be most acutely affected. India should urgently undertake a detailed study of the problem, without waiting for any collaborative endeavor but earnestly trying to meet the formidable challenges of the entire GBM basin, irrespective of the political boundaries, because this is the way environmental impact will occur. It could be an excellent way of promoting collaborative water management in the GBM basin.

We consider development of GBM not in terms of current transboundary conflicts, but with the simple perception that in environmental matters, there is no transboundary and this transcends economic and cultural domain. GBM is a physical, environmental, and historical–cultural entity that has a tremendous challenge to achieve, not only for survival, but attain its due place in the human community, which has to be pursued with utmost urgency and commitment. It is important for all in the GBM basin to realize that they represent the poorest of the poor of the world, as Mahatma Gandhi taught. Not everybody can attain Mahatma Gandhi's capability to identify with the poor, but the message was well accepted by the people and it was this commitment by all and sundry, which led to the achievement of the political Independence. The achievement from the abysmal poverty will also be attained only when this intellectual–emotional commitment is obtained by the people at large. Examination of the technological aspects of development further reinforces the need for an integrated GBM basin approach, neglecting the artificial political boundaries imposed on the environment.

It is easier said than done. However, with the modern state of study of the subject, it is possible to back up the statements with detailed studies in terms of the SESM, supported by transparent participatory models. We can demonstrate that the proposed developmental objectives, technological objectives, and the novel technologies offer the opportunity of revolutionizing the development of water in each characteristic geophysical region, which is almost synonymous with the country and basin, leading to integrated development of the GBM, transcending the boundary issues. There are difficult socioeconomic–institutional–environmental problems, but the challenges may inspire meaningful action if the futures scenes are demonstrated through participatory models.

6.16 River Basin and National Water Planning

The ideas presented are only conceptual. With the modern advances in the science of water resources development and management and the information technology, it is possible to delineate in very detailed scale. The advances have been brought out comprehensively in Chaturvedi (2011a), further advancing the earlier science as brought out in Chaturvedi (1987). The perspective of development of India's waters and the GBM basin has been brought out in detail in Chaturvedi (2011b,c,d). Advances in planning were discussed briefly in Chapter 5 and may be briefly referred to again.

Concepts of national water planning have emerged from time to time. It is interesting to note that Cotton, the most eminent early scientist of India's waters, had proposed a countrywide canal due to his enthusiasm for navigation, even proposing "Communication between India and China by the Line of Burahmpooter and Yang-tsze" (Cotton 1874). It has been followed by K.L. Rao's National Water Grid and currently the National Perspective Plan.

It is time we undertake scientific study of the water resources of India. River basin development is the first consideration. If interlinking is desirable, it will automatically pop up, as the proposed studies indicate.

6.17 Further Advances—Sustainable Development and Water

In view of the large-scale societal development, the subject of sustainability, which has particular focus on management of water, has become a matter of great concern. Much work has been done in the area. Reference to one important study "Our Common Journey—A Transition Toward Sustainability" by the Board of Sustainable Development, a voluntary association of concerned scientists of the US National Reseach Council, is briefly presented, as it may help us in our thinking and action on the subject of management of India's waters (BSD 1999). The focus of the study was global human community and environment. It may be added that India constitutes an overwhelmingly large component of humanity.

The study

> is an attempt to reinvigorate the essential strategic connections between scientific research, technological development, and societies efforts to achieve environmentally sustainable improvement in human well-being. To that end, the Board seeks to illuminate critical challenges and opportunities that might be encountered in serious efforts to pursue goals of development . . .
>
> In the Board's judgment, the primary goals of transition toward sustainability over the next two generations should be to meet the needs of a much larger but stabilizing human population, to sustain the life support systems of the planet, and to susbstantially reduce hunger and poverty . . .
>
> Certain current trends of population and habitation, wealth and consumption, technology and work, connectedness and diversity, and environmental change are likely to persist well into the coming century and could substantially undermine the prospets of sustainability. If they do persist, many human needs will not be met, life support systems will be dangerously degraded, and the numbers of hungry and poor will increase . . . .
>
> Even the most alarming current trends, however, may experience transitions that enhance the prospects of sustainability . . .
>
> Although the future is unknowable, based on our analysis of persistent trends and plausible futures, the Board believes that a successful transition toward sustainability is possible over the next two generations. This transition could be achieved without miraculous technologies or drastic transformations of human societies. What will be required, however,

are significant advances in basic knowledge, in the social capacity and technological capacities to utilize it, and in the political will to turn this knowledge and know-how into action . . .

The Board concludes that most of the individual environmental problems that have occupied most of the world's attention to date are unlikely in themselves to prevent substantial progress in a transition toward sustainability over the next two generations. Over longer time periods, unmitigated expansion of even these individual problems could certainly pose serious threats to people and the planet's life support systems. Even more troubling in the medium term, however, are the environmental threats arising from multiple, cumulative, and interactive stresses, driven by a variety of human activities. Dealing an integrated and place-based understanding of such threats and the options dealing with them is a central challenge for promoting transition toward sustainability . . .

Because the pathway to sustainability cannot be chartered in advance, it will have to be navigated through trial and error and conscious experimentation. The urgent need is to design strategies and institutions that can better integrate incomplete knowledge with experimental action into programs of adaptive management and social learning . . .

Priorities for research in sustainability science and priorities for action in knowledge–action collaboratives were recommended. The Board particularly proposed "integrated approaches to research and actions at the regional scale related to water, atmosphere and climate, and species and ecosystems." We have proposed several conceptual, scientific, and technological revolutions, which are detailed in the companion study of the GBM basin, as an application of the ideas is also demonstrated.

6.18 Conclusion

A revolution in the development and management of water resources of India, be it in the GBM basin, Indus basin, or in the peninsular region, can clearly be established. We can increase the availability of water by an order of magnitude and put the water in the hands of the farmer. We can re-establish free-flowing rivers all over the country. We can ensure good-quality groundwater. Management of floods and the serious impact of climate change remain very serious problems, requiring urgent study. However, the subject needs much scientific study. Even more importunately, vast advances in management of the environment are being made in view of the increasing concern regarding sustainable development (Chaturvedi 2011a). A new perspective of water management has developed, which is particularly important for India because of the tremendous socioeconomic development that will have to be undertaken and the environmental-climatic characteristics.

Notes

1. Collaboration was developed between water planners of the Indus basin and the Ganga basin from the beginning, later leading to a common Board of Consultants, with Dr. A.N. Khosla as the Chairman and Dr. A.C. Mitra as the Vice-Chairman. The author started as a young engineer at Dr. A.C. Mitra's feet in 1946, later joining them as Member in the Board of Consultants.

2. Pfister and Messerli 1990. Some personal observations from the author's visits to Switzerland and Sweden may illustrate the analysis. In one of the lecture-visits to the Swiss Institute of Technology, Zurich, as the author genuinely conveyed his appreciation of the beautiful view of the rich nature and society from the host's window, the host told the author that the Swiss people had petitioned Ferdinand the Great in the tenth century that they be allowed to move over to Sicily, where he had established the capital of his empire. Second, he was surprised to see a large enrollment in the water resources area. He was told that they obtain good employment in the consulting firms working for the developing countries in the design and construction. The author visited the offices of these firms. The same situation was encountered in Sweden, where the author spent several months at a research institute for collaborative research. It was clear that hydro-development had contributed significantly to the economic development and attendant environmental enhancement. In contrast, as the author traveled on foot along the length and breadth of the Himalayas in India, it was painful to see how poverty has ruined the environment and life of the people.

3. The author was posted in Himachal Pradesh as a young engineer in the early 1950s and had the opportunity of studying these developments.

4. The author learned this simple fact incidentally as he was designing the Ramganga project, as Director (Designs), which is the first storage dam to be constructed in the GBM Himalayan region. Soil conservation is usually neglected, as the dam occupies the focus of development. The author calculated the cost–benefit ratio of soil conservation and was surprised to find that it exceeds the cost–benefit ratio of the storage project by several orders of magnitude. The fact was reported to the senior engineers, but soil conservation lost. Later, as he joined the Himalayan rivers development as Member in the Board of Consultants, he would emphasize soil conservation, first and foremost. However, other than receiving an appreciative acknowledgment, he always lost out as far as implementation was concerned.

5. Jagat Mehta, one-time Foreign Secretary, Government of India, led the Indian team in the India–Bangladesh discussions.

6. Crow 1995.

7. The studies were undertaken to educate Indian scientists and engineers on the modern advances in water resources systems planning as sponsored by the Ford Foundation, which was actively promoting modernization of the sector.

8. A study of the possible global change in terms of SESM approach has been carried out by Rothmans and de Vries (1997). A study of future implications on the same lines, focusing on India, was prepared by the author. For some reason, it could not be undertaken. There are current attempts to revive it.

7

Some Perspectives and Institutional and Cultural Revolution

7.1 Introduction

It has been demonstrated that development and management of the water resources in India have to be revolutionized urgently. Considerable potential enhancement of water availability and of advancement exists from technological considerations. However, the management of water over the entire cycle—development, use, and management of the socioeconomic–environmental system—is a societal management issue, above and beyond the proposed technological revolutions. We address these issues. We first briefly restate the current state of water management in India. We analyze the institutional and cultural factors that have led to this phenomenal neglect. We also offer some solutions.

7.2 Current Scene of Water Management in India

Everybody is aware of the poor conditions of water availability and management in the society. The XIth Plan states the poor conditions at the outset. They may be restated in the words of Iyer (2008).[1]

> Taking *urban water supply* first, the general experience in most of the cities is one of a limited, intermittent, unreliable supply; poor water qualty; an unresponsive administration; an inequitable distribution of the available over different areas varying from as low as 30 to 40 litres per capita per day (lpcd) in the areas of poor to more than 400 lpcd in the affluent areas; an implicit subsidation of the rich through low water rates; and an inadequate coverage of the poor by the public system, forcing to buy water at much higher rates from private sources.
>
> Turning to *rural water supply*, despite five decades of planning and more than a decade of "Drinking Water Missions," the curious (and by

now the familiar) fact is that targets for covering "uncovered villages" are repeatedly achieved, but the numbers grow larger rather than smaller. This must mean that some "covered" villages are lapsing back into the uncovered category, and that newer villages are being added to this class. A significant aspect of the scarcity of water in rural areas is of course that the burden of bringing water from distant sources falls on women, including girl children.

Canal water for irrigation (from major and medium projects) is cheap but unreliable. The supply is generally not provided in time and not in quantities needed. The systems in many cases are in disarray because of poor maintenance and operation. The farmer is dependent on irrigation bureaucracy which is not service-oriented, and is (with honorable exceptions) unresponsive to the needs and problems of the water user. The problems of tail-end farmers in the command getting very little water is well known. The system is also amenable to manipulation and distortion by the influence of the rich and the politically powerful, and corruption plays an important role here as elsewhere. The problem of waterlogging and salinity that has been a concomitant of canal irrrigation in many cases is a serious one, but it is primarily a water management than 'governance' issue, though there may be "governance" issues to it. Similarly, in the context of canal waters and irrigation, inter-state river-water disputes have been very prominent but these again are political and water-use than governance issues. . .

In relation to *groundwater* "governance" seems to be non-existent. Both law and politics are to blame here. By law, the water under a piece of land belongs to the owner of that, and he or she (including the corporate entities) can exploit it at will. This could lead to inequitable relations between the seller and buyer of water, the depletion or contamination of aquifer and the drying up of wells and other water-sources in nearby areas. The existing legal position makes regulation very difficult and this is compounded by political factors. Despite the existence of the Central Groundwater Authority for about a decade, and some attempts at legislation at the state level, there is no real regulation of groundwater. . . .

In the context of large projects, "governance" usually presents a high-handed violent and cruel aspect to those who face displacement/loss of livelihood and delayed and badly flawed rehabilitation. To them, "governance" often means the police. The social activists and non-governmental organizations (NGO) that seek to mediate between these project-affected people and the government are likely to perceive "governance" in terms of callousness to suffering, denial of human rights, repression, the use of force in response to resistance, and so on. Those concerned with environmental issues tend to fault "governance" on violations of environmental laws, rules and procedures, and will find serious limitations and inadequacies in Environmental Impact Assessments (EIAs).

From the financial, economic and management points of view, low water rates and consequently ills of poor revenues, loss in financial terms, inadequacy of funds even for operation and maintenance (O&M) and non-availabilty of funds for capital-renewal and new investment, will constitute bad governance.

From the point of view of governmental agencies concerned with water—the municipal authorities and the irrigation Departments—proper governance will undoubtedly seem hamstrung by a chronic inadequacy of budgetary allocations; a shortage of water (in many cases) for the services that they are expected to provide; the pressures of the rich and the politically well-connected; a powerlessness to deal effectively with those who violate the law or bend or circumvent the procedure; improper directions from the political levels; and the pervasive presence of corruption. (The reference is to the perceptions of the upright and conscientious officials; others may prefer to conform to the prevailing ethos and make their "adjustments.")

The various water governance problems and deficiencies enumerated above are usually seen piecemeal, and many partial remedies and fragmented reforms are proposed: Participatory Irrigation Management or PIM; privitazition; water markets; proper pricing of water; and so on. Some of these "reforms" originate in the government, some are recommendations of the World Bank and International Monetary Fund (IMF), and some are ideas advocated by our own economists. Some of these propositions are sound, some are questionable and some are fraught with danger.

From the world of environmentalists and social activists and mobilizers there are pleas for a National Rehabilitation Policy (now in place, and as mentioned above, a disappointment); the drastic modification of the Official Secrets Act and the enactment (or full implementation) of a Freedom for Information Act; a thorough going reform of the Land Acquisition Act; full adherence to the Provision of the Panchayats (Extension to the Scheduled Areas) Act 1996; due conformity to the provisions of the Environment Protection Act; proper implementation of the requirement, now mandatory, of public hearing in the case of large projects; and so on.

Reforms in the area are indeed very necessary. However, if we consider any "water governance" issue carefully, we will find ourselves led beyond governance in a narrow sense into larger issues, and beyond the sphere of governance into domains of water users, private sector agencies, and civil society.

7.3 Challenge of Management of Water

As Iyer concludes, we are led to larger issues if we consider the subject of governance, or more appropriately, management of water. Water has a life cycle of development, use, and return flows. It has to be managed efficiently at each of these life cycles. At each, it is integrated with the management of the corresponding socioeconomic activity. We may look at each of the socioeconomic activities and the conjunctive management of water.

First is the issue of managing drinking water and the return flows. There are two diffferent worlds: rural and urban. The rural world represents the overwhelming proportion of India. It is terribly undeveloped and indifferent to drinking water issues. A personal experience will bring it out significantly.

The author, on his return from higher studies in the United States, joined Indian Institutes of Technology, Kanpur (IITK), as the Founding Head, Department of Civil Engineering. IITK was being established with the support of nine U.S. universities. The campus and housing were still being developed. The U.S. faculty and some senior Indian faculty, the author being one of them, lived on the campus housing. A U.S. faculty member suggested that according to the U.S. practice, we should find out what the community expects from us. Accordingly, we, including Dr. (Mrs.) Vipula Chaturvedi, who was also on the faculty, went to the neighboring village and met the village chiefs, who were enjoying the evening with a *hukka* (Indian smoking device) round the *chaupal* (meeting place). On asking them as to what they wanted, the *pradhan's* (chief) immediate response was "water." Our query was "for drinking purposes?" The pradhan responded, "Oh no, that is brought by the girls from the village pond. The land wants water for all it gives us." Vipula asked the women folk, "Should not the girls be going to school?" The response was that their life is meant for serving the men folk. On learning that she was teaching at the IITK, they were horrified. The men folk giggled, "What sort of men are they, being taught by women!" Several villages were visited. The response was almost identical. No wonder, most of rural India is not supplied with water for drinking and domestic purposes. There is neither the demand nor is there any interest by the supplying officials. From a technological perspective, there is no problem in adequately managing water at the rural level. The challenge is at an entirely different level of rural development.

The urban world story is equally frustrating. Given the example of New Delhi, which is the capital (and where the author lives), the scene was described in the opening pages of the book. Water is not supplied in adequte quantity, or according to needs, and it is not potable. The scene is even worse in the urban world all over India. Again, the people are not aware of the terrible deprivation, and, therefore, there was no pressure coming from them on the civil authorities to supply them domestic water decently according to developed countries. As far as the civic authorities are concerned, their focus is on making illegal money. Urban development is a very complex issue, which has been discussed elsewhere (Chaturvedi and Chaturvedi 2012).

Agriculture requires the dominant share. Agricultural supplies are an apology of adequate and timely requirements. No wonder, yields in Indian agriculture are one of the lowest even among the developing countries.

The subject does not end at providing water to different agencies for certain activities, as is conventionally done, but at managing water over the entire life cycle in the context of these activities, as emphasized earlier. For instance, even before supplying water for different demands such as for drinking purposes or irrigation, water has to be managed at the watershed level. Next, beyond supplying water to agencies for urban or rural users, the return flows have to be managed. That is a simple matter as far as technology is concerned, but the subject gets integrated with urban or rural management.

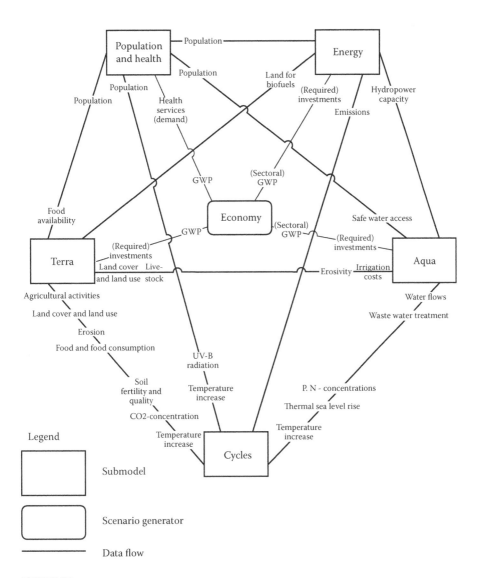

FIGURE 7.1
Societal–environmental systems management. (From Chaturvedi, M.C., *Environmental Systems Management for Sustainable Development*, in press, 2011a.)

Similarly, water has to be managed over the life cycle in agriculture, industry, and the natural state of the environment itself. Figure 7.1 (Figure 6.11 reproduced) schematically illustrates the dynamics of the Societal Environmental Systems Management in which water is embedded (Rothmans and de Vries 1997; Chaturvedi 2011a). In considering the management of water, it was emphasized that it has to be considered over this cycle. However, we are going beyond water for the management of the different components of the socioeconomic system.

7.4 Paradox

However, it is mandatory. It has been discussed in detail by Chaturvedi (2011a) in this perspective, to which reference has been made. This is particularly important for India in view of the large-scale development of the economy and the environment. The NCIWRD 1999 also referred to the environment, though in a very limited perspective. Yet, almost no action has been taken. The natural reaction is to ask why there is this callous indifference.

Ultimately, it is the society that is responsible for the state of social affairs. An attempt has, therefore, been made to identify the social scene. This is based on the vast experience of the author over his life and in the field of water. It is, therefore, presented in the form of a narrative statement (Appendix 2). The conclusions arrived at are reproduced.

> The review shows that the society is not conscious of the possibilities of the development that humanity has achieved. This is not realized even in basic terms such as better living conditions, easy availability of potable water, proper sanitation facilities, transport, and, above all, basic education. The prejudices against the female members of the society are still those of primitive conditions. Even in economic activities, people are not conscious that their well-being can be dramatically increased. Agriculture, on which the vast majority depends, can be transformed if science and technology is brought to bear on it, in which scientific use of water is crucial. They cannot be expected to even visualize the danger that the society faces if urgent rapid development of the society is not undertaken, which becomes more threatening with the ongoing climate change.
>
> The managers of the society, political and bureaucratic, except for some honorable ones, are totally indifferent to the people's concerns or future challenges, even with the historic precepts immortalized in religion and, recently, in the person of Mahatma Gandhi. Generally (but for honorable exceptions), they are engrossed in seeking personal benefits and not even committed to public duty.
>
> Landes, an eminent economic historian, has made an exhaustive study of the development of the Western world, bringing out the important role that technology has played (Landes 1969). He followed the study to explore why some countries are so rich and some are so poor (Landes 1998). After a monumental study, he came to a simple conclusion: culture makes all the difference (Landes 1998, 2000).

A personal observation may be reproduced to emphasize this point. The proposed Chaturvedi Water Power Machine will make dramatic improvements in the development of water. It has received an encouraging response from the highest levels. It occurred to the author that while this will take time, he asked why we should not start with some projects that can be undertaken immediately. He phoned one of his former students in Uttar Pradesh, who holds one of the most senior positions. His response was "Sir, have you not read today's papers about the murder of one young officer who refused to give his quota to the Chief Minister's birthday celebrations?[3] The objective function here is making money for our political masters and us, not engineering."

7.5 Background of Water Management in India

In the British period, India was ruled by Her Majesty's Government. It was managed from the perspective of maintaining the political control and maintaining law and order, just as the earlier Mughal Emperors had ruled, except that the Emperor or Empress, who was white, was set in throne far away in an island and ruled India through a handful of her white British officials. In the early phases of the British rule, most of the activities were undertaken by the army. Gradually, a civil service was instituted. The senior positions were held by British officers. They engaged a vast number of local people in subordinate positions to do their command. India was a unique Colonial Government. People accepted the humiliating subjugation for a long time.

Water resources development also started in this perspective. The development and management were undertaken by the military officers, and gradually a civil service was established. The officers came from England, but as the canals were constructed, Indians were recruited as junior officers.

There was hardly any industrial development or urbanization in the British period. Development of water was only in the context of irrigation, and water resources development became synonymous with irrigation. Irrigation meant providing an apology of water requirements to stabilize the sustenance agriculture. Even for irrigation, the colonial government was content with supplying water up to the canal outlet for Rabi (winter period in local parlance), leaving the rest of the task of conveyance and distribution to the farmers themselves. Activities such as land-leveling, drainage, lining of water courses, and so on, which demanded public participation, were also not considered possible by a foreign government. They were not within the capability of the poor peasantry and were therefore neglected. Thus, in short, a policy of inefficient extensive irrigation through low-flow diversions to stabilize the sustenance agriculture evolved from a convergence of social, political, economic, technological, and financial considerations.

A new era, expressed by Nehru's tryst with destiny speech in the Parliament on the occasion of achievement of Independence, started after Independence in 1947. Planned development was undertaken with the First Five-Year Plan (1951–1956), and water resources development was given the highest importance. Water resources development, however, was synonymous with irrigation and multipurpose projects.

There has been an impressive development of water resources in India since Independence, but the concept and the institutional setup, though expanded enormously and even transformed, have maintained the colonial ideas, objectives, and most importantly, "culture." Irrigation, however, has been undertaken with total disregard to productivity and environmental considerations. India has one of the largest irrigation systems of the world, but also one of the poorest in terms of productivity and environmental conservation, as brought out earlier. Kharif irrigation is still almost neglected, except for some minor developments. Large-scale water withdrawals from rivers and polluted inflows have turned rivers into fetid sewers over large stretches, in view of the increasing urbanization and industrialization, leading to a serious problem of water supply to urban habitats/residents. There is an increasing problem of groundwater overexploitation and pollution. And yet, in commissions and in private studies, this basic fact has been totally ignored. Of course, the official agencies are totally indifferent about the utterly poor system, much less having any knowledge of the scientific revolution that has taken place in the field of environmental systems management, of which water is a vector. Their objective, as that of their political bosses, is making money. This is not a reflection on the management of water: it is true that all sectors of government are involved in public affairs.

7.6 Institutional Revolution

The historic perception of management of water was irrigation, and institutional arrangements got organized in this context, particulaly in the context of management of water in the Indus basin in the British India (Gulhati 1972). Water continued to be seen in terms of irrigation, and even the Ministry at the Government of India level was named the Ministry of Irrigation. It has been renamed the Ministry of Water Resources, but the perspective has not changed much. The change has been even less at the state level, where actual activity resides. A fundamental issue is the conceptual and, correspondingly, institutional revolution. What form it has to take is a complex issue as it is linked with institutional administrative issues in a wider context. It is, therefore, not possible to make any recommendations at this stage.

7.7 Transformation

Besides institutional reformation, there is the issue of culture. This is also a very complex issue as we are talking about culture in public management. Commissions have been appointed from time to time. They have given their recommendations, for all their worth. These were relegated to the waste paper basket. The current study will have a similar fate.

The issue is what has come to be called a matter of "culture." A new culture has to be introduced. We will not venture as to how the revolution in the system can be brought about because it is not a matter of water alone. Perhaps some impact could be made if a small action-research group of academicians and professionals, who may be interested in revolutionizing the setup, undertakes modernization of the system in scientific–environmental–technical–managerial terms by looking at the various aspects of development. Perhaps the effort may snowball. It may be emphasized that the academic community, teachers and students alike, was intimately involved in the freedom movement.[4] This has to be repeated in India's fight against poverty and attainment of the due global standard. An attempt has been made by the author to initiate the activity.[5]

7.8 Involvement of Foreign Agencies

It is necessary and important to emphasize that the development should be undertaken by the developing countries individually and collaboratively by themselves. It is also important that it is undertaken on equal terms. India has to be particularly sensitive to this fact.

Involvement of foreign agencies is encouraged by international lending and donor agencies. This should be specifically ruled out. It is wrong, in principle, as it kills development of indigenous initiative and capability. All developed countries achieved development on their own. Furthermore, history is testimony to the fact that involvement of foreign powers is suicidal. Also, it should be noted that the foreign scientists have little technical experience of Indian problems. Knowledgeable U.S. scientists have acknowledged the fact and desisted from undertaking studies of Indian river basins (Bower and Hufschmidt 1984). The limitations of a recent study by Briscoe and Malik 2006, as discussed earlier, confirm their and our perceptions. A review will bring out that it totally misses the advancements we have proposed.

To give another example, scientific analysis has convincingly demonstrated that at our stage of development and in our physiographic–hydrologic conditions, dams are essential, but foreign scientists, supported by the World

Bank, were actively against them (Chaturvedi 2011c,d). The Government of India has already precluded working of foreign agencies in the GBM basin.

Notes

1. Iyer 2008. R.R. Iyer was the Secretary, Ministry of Water Resources, Government of India, from 1985 to 1987.
2. NCIWRD 1999, p. 1.
3. The MLA has been awarded life imprisonment, *Hindustan Times*, New Delhi, Saturday, May 7, 2011, p. 10.
4. The author's first experience as he joined Allahabad University in 1942 as a freshman was attending a public lecture by Jawaharlal Nehru on August 6, 1942, before he left for the historic August 9 Congress meeting where the Quit India resolution and implementation were undertaken. As the leaders were imprisoned, demonstrations were held all over the country, including in the author's university, in which all joined. In the police firing, many students were killed, including an author's class fellow.
5. The proposal was made to the Honorable Union Minister that academics of the five IITs in the GBM basin and concerned official professionals undertake a collaborative study of the challenging problems of the basin. According to his instructions, a group is trying to be constituted by representatives of the profession and academia. It is hoped that the study will result in some action. It is a matter that cannot be given up because, otherwise, we are doomed. In our judgment, it is a basic duty of the academics to watch the interest of the society. In this context, an important development has taken place. National Ganga River Basin Authority has been established under the Ministry of Environment and Forests on February 20, 2009. On reading the news in the papers, the author wrote to the Honorable Minister, Shri Jairam Ramesh, proposing that the planning be undertaken collaboratively by the IITs and the Ministry of Environment and Forests. This was accepted and a Memorandum of Understanding between the two agencies for the preparation of the Ganga River Basin Management Plan has been arrived at on July 6, 2010.

8

Conclusion

India and China are at the bottom of human community in terms of the socioeconomic state, which is almost like a wine glass, as shown in Figure 8.1. They constitute the dominant component of the backward community of humanity, each about 20%. India faces an enormous challenge of development. Management of the environment will be one of the most important components of the activities because in the last analysis, humanity depends on the environment. Indeed, society and the environment are one integral component, as the Indian sages discovered long ago and as modern science has rediscovered. An integrated societal–environmental systems management has to be undertaken, as modern science puts it. Advances have been made in the area leading to new concepts, technologies, policies, and management (Chaturvedi 2011a). The current activities have to be revolutionized in these terms.

The subject is a matter of tremendous importance and urgency. The currently industrialized countries had the whole world at their disposal. The developing countries, in which China and India are the dominant components, have limited environmental resources. For example, focusing on water, two characteristics stand out. First, the per capita availability is comparatively very limited. For instance, the figures of per capita availability of freshwater resources (cubic meters per capita) for India and the United States are 2167 and 9259, respectively. In addition, in view of the climatic–hydrologic characteristics, the development of water for agriculture is a central challenge, requiring an overwhelming component of the development of water for meeting the agriculture requirements in India, which accounts currently for about 80% of the total water use; not much such effort is required in large parts of the industrialized world of Europe and the United States, and their requirement for the agricultural sector is only about 10% of the total development. Serious environmental strains are already being felt, while there is still a very long road of development to be traversed. Creativity and science are, therefore, essential and urgent.

Environmental management requires integrated management of all its components, besides the integration with the society. Water is an important component. Scientific development of the waters of India was briefly presented, based on several detailed studies (Chaturvedi 2011a,b,c,d). Reference may also be made to fundamental advancements in concepts, which have been detailed in the companion study of Ganga–Brahmaputra–Meghna (GBM) waters, as it provided an opportunity of application. It has been demonstrated that a

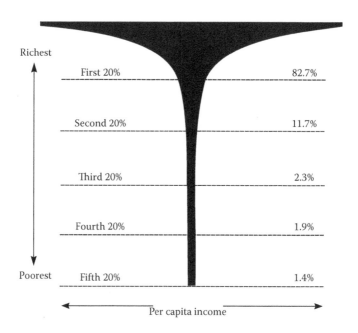

FIGURE 8.1
Distribution of world income and disparities.

revolution is urgently needed from scientific, technological, environmental, and social considerations in concepts, technologies, and management. It has been demonstrated that it is eminently possible to manage the waters of India satisfactorily, despite the enormous challenges.

A revolutionary advancement was proposed in terms of concepts and, in turn, in terms of policies. Paradoxically, following from these considerations, some novel technologies emerged, which revolutionize the water resources development of India. Focusing on the GBM basin, which is the centerpiece of India's waters, it was brought out that, taking into account its unique physiographic and hydrologic characteristics, the perspective in the Himalayan regions is to utilize the enormous energy potential, instead of focusing on the low flows, as being done currently. A new policy and technology called Chaturvedi Water Power Machine emerges. It is proposed that, in contrast to the current policy, the low flows should be left untouched to meet the environmental and domestic demands. The energy of the high flows should be utilized by pumping them up for storage in the Vindhyan region and for release later in terms of pumped storage. The contrast with the current conventional pumped storage is that we are envisaging intertemporal, that is, over the year, and interspatial release. The releases are on the other side of the Vindhyas, leading to transfer to the rest of India. As stated earlier, we are killing five birds with one stone: (1) environmental conservation, (2) large increase in water, (3) large increase in peaking energy, (4) intertemporal–interspatial pumped storage, and (5) most economical interbasin transfer.

Similarly, another revolutionary policy and technology, called Chaturvedi Ganga Water Machine, emerges, as one considers management of the Gangetic alluvial plains. It is proposed that the focus should be on storage of water on the alluvial plains, concurrently with its use. This readily follows if we focus on the management of the surface water on the watershed and Kharif (monsoon period) irrigation, on one hand, and Rabi (winter period) irrigation through groundwater, on the other hand, managed conjunctively.

It is proposed that Rabi irrigation is undertaken through groundwater, conjunctively lowering it to precalculated levels so that it facilitates the groundwater recharge during the monsoon period through careful management of the surface waters and Kharif irrigation. The low flows should be conserved from environmental and domestic demand considerations.

Another novel technology emerges—Chaturvedi Reversible Pumps. The pumps, which are proposed to be used essentially during Rabi, are made reversible, enabling recharge of the groundwater during the monsoon period.

Some other novel possibilities also emerge. There is a possibility of artesian waters in the GBM basin, which is somehow ignored for some unknown reasons. This should be explored.

It was demonstrated that the water availability in India can be increased dramatically from the current estimates of 1032.75 to 1799.89 km^3, representing an increase of about 180%. The availability in the peninsular region can also be increased dramatically to a figure of about 483.15 km^3 from the current estimated availability of about 298.13 km^3, representing an increase of about 230%. Moreover, with the large-scale surface, groundwater, and hydro energy development, there is a possibility of scientific surface, groundwater, and power planning using modern scientific systems management developments.

Besides advancements in enhancing the water availability, emphasis has also to be laid equally on their efficient management for use and the return flows over the hydrological cycle. The productivity in agriculture has to be increased to the currently achieved levels internationally, and the quality and quantity of the return flows in the industrial and domestic cycle have to be improved.

All this, however, depends on the state of the society. A paradoxical conclusion emerges. Rapid socioeconomic development is essential for scientific management, including conservation, of the environment and vice versa. Thus, the issue is not of novel technologies and vastly increased availabilities of water, but of new attitudes and concepts. The socioeconomic–environmental revolution has to be set about urgently. The challenge has been well expressed by Amartya Sen. Studying India's development, he concluded, "It is said that the British Empire was founded in a state of absent mindedness. It is unlikely that a happy and prosperous India can be built that way" (Sen 1989).

References

Chapter 1

Chaturvedi, M.C. 1976. *Water—Second India Studies.* New Delhi: Macmillan.

Chaturvedi, M.C. 2011a. *Societal Environmental Systems Management* (in press).

Chaturvedi, M.C. 2011b. *India's Waters—Environment, Economy and Development.* Boca Raton, FL: Taylor & Francis.

Chaturvedi, M.C. 2011c. *India's Waters—Advances in Development and Management.* Boca Raton, FL: Taylor & Francis.

Chaturvedi, M.C. 2011d. *Ganga–Brahmaputra–Meghana Waters—Development and Management.* Boca Raton, FL: Taylor & Francis.

Chaturvedi, M.C. and P. Rogers (eds.). 1985. *Water Resources Systems Planning–Some Case Studies for India.* Bangalore, Indian: Academy of Sciences.

Cline, W.R. 2007. Global warming losers. *The International Economy*, 21(4), Fall 2007.

Kumar, M.D. 2010. *Managing Water in River Basins.* New Delhi: Oxford University Press.

National Commission for Integrated Water Resources Development (NCIWRD). 1999. Report, Vol. I, Integrated Water Resources Development—A Plan for Action. Ministry of Water Resources, Government of India, New Delhi.

Pasucal, U., A. Shah, and J. Bandopadhyay (eds.). 2010. *Water, Agriculture and Sustainable Well-being.* New Delhi: Oxford University Press.

Rogers, P. 2008. Facing the freshwater crisis. *Scientific American*, August 2008, 28–35.

Rogers, P. (ed.). 2011. Treatise on Water Science, Vol. 1, *Management of Water Resources.* Elsevier: Amsterdam.

Sen, A. 2009. We have reason to be ashamed. *Tehelka*, January 2009, 36–38.

Vaidyanathan, A. 2010. *India's Water Resources.* New Delhi: Oxford University Press.

World Resources Institute. 1996. *World Resources 1996–97.* New York: Oxford University Press.

Chapter 2

Bruinzeel. 1989. pp. 4 and 5.

Chaturvedi, M.C. 1976. *Water—Second India Studies.* New Delhi: Macmillan.

Chaturvedi, M.C. 2001. Sustainable development of India's waters—Some policy issues. *Water Policy* 3, 297–320.

Chaturvedi, M.C. 2011. *India's Waters—Environment, Economy and Development.* Boca Raton, FL: Taylor & Francis.

Government of India (GOI). 1972. Report of the Second Irrigation Commission. Ministry of Irrigation, New Delhi.

National Commission for Integrated Water Resources Development (NCIWRD). 1999. Report, Vol. I, Integrated Water Resources Development—A Plan for Action. Ministry of Water Resources, Government of India, New Delhi.

World Bank. 1997. *World Resources 1996–97*. New York: Oxford University Press.

World Resources Institute. 1996. *World Resources 1996–97*. New York: Oxford University Press.

Chapter 3

Buckley, R.B. 1893. *The Irrigation Works of India*, 42nd edn. London: E & FN Spon.

Chaturvedi, M.C. 1973. Indian National Water Plan and Grid, First World Congress, International Water Resources Association, Chicago, Sept. 24–28.

Government of India (GOI). 1972. Report of the Second Irrigation Commission. Ministry of Irrigation, New Delhi.

Gyawali, D. 2001. *Water in Nepal*, p. 137. Nepal: Himal Books.

Madison, A. 1971. *Class Structure and Economic Growth: India and Pakistan since the Moguls*. London: Allen and Unwin.

National Commission for Integrated Water Resources Development (NCIWRD). 1999. Report, Vol. I, Integrated Water Resources Development—A Plan for Action. Ministry of Water Resources, Government of India, New Delhi.

Schwarz, H.E., J. Emel, W.J. Dickens, P. Rogers, and J. Thomson. 1990. Water quality and flows, in *The Earth as Transformed by Human Action*, edited by B.L. Turner II, W.C. Clark, R.W. Kates, J.E. Richards, and W.B. Myers. Cambridge: Cambridge University Press.

Spear, P. 1978. *A History of India*, Vol. II. New Delhi: Penguin Books.

Thapar, R. 1977. *A History of India*, Vol. I. New Delhi: Penguin Books.

Chapter 4

Briscoe, J. and R.P.S. Malik. 2006. India's Water Economy—Bracing for a Turbulent Future. New Delhi: The World Bank.

Chaturvedi, M.C. 1976. *Water-Second India Studies*. New Delhi: Macmillan.

Chaturvedi, M.C. 1987. *Water Resources Systems Planning and Management*. New Delhi: Tata McGraw-Hill.

Chaturvedi, M.C. 1998. Utilisation of National Waters for Irrigation. Memorandum submitted to the Government of India, Parliamentary Agriculture Standing Committee on Their Invitation, New Delhi.

Chaturvedi, M.C. 2001. Sustainable development of India's waters—Some policy issues. *Water Policy* 3, 297–320.

Chaturvedi, M.C. 2011a. *Environmental Systems Management for Sustainable Development* (in press).

Chaturvedi, M.C. 2011b. *India's Waters—Environment, Economy and Development*. Boca Raton, FL: Taylor & Francis.

Chaturvedi, M.C. 2011c. *India's Waters—Perspectives in Development*. Boca Raton, FL: Taylor & Francis.

Chaturvedi, M.C. and V.K. Srivastava. 1979. Induced ground water recharge. *Water Resources Research* 15, 1156–1166.

Chaturvedi, M.C., R. Revelle, and V.K. Srivastava. 1975. Ganga Water Machine II. Induced Ground Water Recharge. Second World Congress. International Water Resources Association. New Delhi.

Conway, G. 1998. *The Doubly Green Revolution*. Ithaca, NY: Cornell University Press.

Government of India (GOI). 1972. Report of the Second Irrigation Commission. Ministry of Irrigation, New Delhi.

India Water Partnership (IWP). 2000. India Water Vision 2025—Report of the Vision Development Consultation. India Water Partnership, New Delhi Institute for Human Development.

Iyer, R.R. 2003. *Water: Perspectives, Issues, Concerns*. New Delhi: Sage Publications.

Iyer, R.R. 2007. *Towards Water Wisdom: Limits, Justice, Harmony*. New Delhi: Sage Publications.

Iyer, R.R. 2008. *Governance of Water*, edited by V. Ballabh. New Delhi: Sage Publications.

Mohile, A.D. 1996. Interbasin transfers of water for national development—Problems and prospects. Theme Paper. Water Resources Day—1996. Indian Water Resources Society, Roorkee.

National Commission for Integrated Water Resources Development (NCIWRD). 1998. Report of the Working Group on Interlinking of India's Rivers. Ministry of Water Resources, Government of India, New Delhi.

National Commission for Integrated Water Resources Development (NCIWRD). 1999. Integrated Water Resources Development—A Plan for Action, Vol. 1. Ministry of Water Resources, Government of India, New Delhi.

Reddy, M.S. 1991. Inter-Basin Transfers—National Perspectives. 2nd National Water Convention, New Delhi.

Revelle, R. and T. Herman. 1972. Some possibilities for international development of the Ganga basin. Research Report, Harvard University, Centre for Population Studies. MA. USA.

Revelle, R. and V. Lakshminarayan. 1975. Ganges water machine. *Science* 188, 611–616.

Singh, B. 2003. A big dream of little logic. *Hindustan Times*, March 9.

Srivastava, V.K. 1976. Study of Induced Groundwater Recharge, Ph.D. Thesis, Indian Institute of Technology, Delhi, New Delhi.

Task Force on Interlinking of Rivers (TFIR). 2003. Interbasin Water Transfer Proposals. Ministry of Water Resources, Government of India, New Delhi.

Wilderer, P. (Editor-in-Chief) 2011. Treatise in Water, Vol. 1, in *Management of Water Resources*, edited by P. Rogers (Volume Editor). Elsevier: Amsterdam.

World Resources Institute. 1996. *World Resources 1996–97*. New York: Oxford University Press.

Chapter 5

Arya, Y.C. 1980. Dynamic system response of crops for improving management of irrigation. PhD Thesis, Indian Institute of Technology, Delhi, New Delhi.

Arya, Y.C. and M.C. Chaturvedi. 1981. Dynamic system response of crops. *Water Resources Research*.

Bairoch, P. 1988. *Cities and Economic Development—From the Dawn of History to the Present*. Chicago: Chicago University Press.

Bhatia, B.M. 1967. *Famines in India*. Bombay: Asia Publishing House.

Board on Sustainable Development (BSD), National Research Council, 1999. Our Common Journey: Transition toward Sustainability. National Academy Press, Washington, DC.

Bossel, H. 1998. *Earth at Cross Roads: Paths to Sustainable Development*. Cambridge, UK: Cambridge University Press.

Cautley, P.T. 1860. Completion Report. Ganges Canal. London: Smith, Elder & Co.

Central Water Power Commission. 1997. *Storages in River Basins of India*. Government of India, New Delhi.

Chaturvedi, M.C. 1973. Indian National Water Plan and Grid, First World Congress, International Water Resources Association, Chicago, Sept. 24–28.

Chaturvedi, M.C. 1985. The Ganga–Brahmaputra–Barak basin, in *Water Resources Systems Planning—Some Case Studies for India*, edited by M.C. Chaturvedi and P. Rogers. Bangalore: Indian Academy of Sciences.

Chaturvedi, M.C. 1987. *Water Resources Systems Planning and Management*. New Delhi: Tata McGraw-Hill.

Chaturvedi, M.C. 2001. Sustainable development of India's waters—Some policy issues. *Water Policy* 3, 297–320.

Chaturvedi, M.C. 2007. Management of Ganga–Brahmaputra–Meghana Waters—People Oriented Global Leadership Project. Ministry of Water Resources, Government of India.

Chaturvedi, M.C. 2011a. *Environmental Systems Management for Sustainable Development* (in press).

Chaturvedi, M.C. 2011b. *India's Waters—Environment, Economy and Development*. Boca Raton, FL: Taylor & Francis.

Chaturvedi, M.C. 2011c. *India's Waters—Perspectives in Development*. Boca Raton, FL: Taylor & Francis.

Chaturvedi, M.C. 2011d. *Ganga–Brahmaputra–Meghana Waters—Development and Management*. Boca Raton, FL: Taylor & Francis.

Chaturvedi, M.C. and V. Chaturvedi. *India—The Challenge of Development* (in preparation).

Chaturvedi, M.C. and V.K. Srivastava. 1975. Induced ground water recharge. *Water Resources Research*, 15, 1156–1166.

Chaturvedi, M.C., R. Revelle, and V.K. Srivastava. 1975. Ganga water machine II—Induced groundwater recharge. Workshop on Flood Mitigation and Water Resources Development, Second World Congress, International Water Resources Association, New Delhi.

Cline, W.R. 2007. Global warming losers. *The International Economy*, 21(4), Fall 2007.

Frosch, R.A. and N.E. Gallopoulos. 1989. Strategies for manufacturing. *Scientific American*, 261(3): 144–152.

Gandhi, P.J. 2005. *Dr. Kalam's PURA Model and Societal Transformation*. New Delhi: Deep Publications.

Government of India (GOI) 2002. Tenth Five-Year Plan. Planning Commission. New Delhi.

Gosain, A.K., S. Rao, and B. Debajit. 2006. Climate change impact assessment on hydrology of Indian rivers. Current Science Special Section. *Climate Change and India*, 90(3), 10 February 2006.

Hall, W.A. and J.A. Dracup. 1970. *Water Resources System Engineering*. New York: McGraw Hill.

Hasan, Q. 2005. A study of sustainable development of waters of India. PhD Thesis, Indian Institute of Technology, Delhi.

Hoekstra, A.V. 1998. *Perspectives on Water—An Integrated Model-Based Exploration of the Future*. Utrecht, The Netherlands: International Books.

Jones, P.H. 1985. Deep Aquifer Exploration Project—Upper Gangetic Plain, India. Prepared for the World Bank. P.H. Jones Inc.

Khepar, S.D. 1980. System studies for microlevel irrigation water management. PhD Thesis, Indian Institute of Technology, Delhi, New Delhi.

Khepar, S.D. and M.C. Chaturvedi. 1979. *Optimum Decisions for Lining irrigation Canal Distribution Networks*. Netherlands Agricultural Water Management. Elsevier.

Kothari, M. 2000. Sustainable water resources development for a river basin. PhD Thesis, Indian Institute of Technology, Delhi.

Lakshsminarayana, V. and R. Revelle. 1975. Ganga Water Machine. Science 188: 541–549.

Landes, D.S. 1999. The Wealth and Poverty of Nations-why some are rich and some so poor. New York. W. W. Norton.

Mass et al. 1962. *Design of Water Resources Systems*. Cambridge: Mass Harvard University Press.

Mehta, J.S. 1992. *Politics of Riparian Relations*. New Delhi: Centre for Policy Studies.

National Commission for Integrated Water Resources Development (NCIWRD). 1999. Report, Vol. I, Integrated Water Resources Development—A Plan for Action. Ministry of Water Resources, Government of India, New Delhi.

Nehru, J. 1947. The Discovery of India. London. Penguin.

Prasad, R.K. 1981. Finite element simulation of groundwater system. PhD Thesis, Indian Institute of Technology, Delhi, New Delhi.

Rao, K.L. 1975. India's water wealth-its assessment, uses and projections. New Delhi. Orient Longmans.

Revelle, R. and T. Herman. 1972. Some possibilities for international development of the Ganga basin. Research Report, Harvard University, Centre for Population Studies, MA, USA.

Revelle, R. and V. Lakshminarayana. 1975. Ganges water machine. *Science*, 188, 611–616.

Rothmans, J. and B. de Vries (eds.). 1997. *Perspective on Global Change—The TARGETS Approach*. Cambridge: Cambridge University Press.

Srivastava, V.K. 1976. Study of induced groundwater recharge. PhD Thesis, Indian Institute of Technology, Delhi, New Delhi.

Swaminathan, M.S., M.C. Saxena, M.C. Chaturvedi, M. Gopal, and O. Thanvi. 2002. *Lava ka Baas*. New Delhi: Centre for Science and Environment.

Todaro, M.P. 1989. *Economic Development in the Third World*. New York: Longman.

Todaro, M.P. and S.C. Smith. 2003. *Economic Development*. Singapore: Pearson Education.

Waggoner. 1997. How much land can ten billion people spare for nature, in *Technological Trajectories and Human Environment*, edited by H.A. Jesse and H.D. Langford. Washington, DC: National Academy Press.

World Resources Institute. *World Resources 1996–97*. New York: Oxford University Press.

Chapter 6

Ahmad, Q.K., N. Ahmad, and K.B.S. Rasheed (eds.). 1994. *Resources, Environment and Development in Bangladesh: With Particular Emphasis on the Ganges, Brahmaputra and the Meghan Basins*. Dhaka: BUP/Academic Publishers.

Ahmad, Q.K., U.A. Ahsan, H.R. Khan, and K.B.S. Rasheed. 2001a. GBM regional water vision: Bangladesh perspectives, in *Ganges–Brahmaputra–Meghana Region—A Framework for Sustainable Development*, edited by Q.K. Ahmad, A.K. Biswas, R. Rangachari, and M.M. Sainju (eds.). Dhaka: The University Press Ltd.

Ahmad, Q.K., A.K. Biswas, R. Rangachari, and M.M. Sainju (eds.). 2001b. *Ganges–Brahmputra–Meghana Region—A Framework for Sustainable Development*. Dhaka: The University Press Ltd.

Arya, Y.C. 1980. Dynamic system response of crops for improving management of irrigation. PhD Thesis, Indian Institute of Technology, Delhi, New Delhi.

Asthana, B.N. 1984. System studies for regional environmental resources planning in a developing economy. PhD Thesis, Indian Institute of Technology, Delhi, New Delhi.

Bhatia, P.K. 1984. Modeling integrated operation of multipurpose multireservoir water resources systems. PhD Thesis, Indian Institute of Technology, Delhi, New Delhi.

Bower, B.T. and M.N. Hufschmidt. 1984. A conceptual framework for analysis of water resources management in Asia. *Natural Resources Forum*, 8(4), 343–356.

Buckley, R.B. 1880. *The Irrigation Works of India and Their Financial Results*. London: A.H. Allen & Co.

Chaturvedi, M.C. 1973. Indian National Water Plan and Grid, First World Congress, International Water Resources Association, Chicago, Sept. 24–28.

Chaturvedi, M.C. 1987. *Water Resources Systems Planning and Management*. New Delhi: Tata McGraw-Hill.

Chaturvedi, M.C. 2011a. *Environmental Systems Management for Sustainable Development* (in press).

Chaturvedi, M.C. 2011b. *India's Waters-Environment, Economy and Development*. Boca Raton, FL: Taylor and Francis.

Chaturvedi, M.C. 2011c. *India's Waters-Advances in Development and Management*. Boca Raton, FL: Taylor & Francis.

Chaturvedi, M.C. 2011d. *Ganga–Brahmaputra–Meghana Waters—Development and Management*. Boca Raton, FL: Taylor & Francis.

Chaturvedi, M.C. and P. Rogers (eds.). 1985. *Water Resources Systems Planning—Some Case Studies for India*. Bangalore: Indian Academy of Sciences.

Chaube, U.C. 1983. Two Level Multi-Objective Reconnaissance Study of a Large Water resources System. Ph.D. Thesis. Indian Institute of Technology, Delhi.

Cotton, Lt. Gen. Sir Arthur. 1874. Lectures on Irrigation Works in India. Delivered at the School of Military Engineering, Chatham, and Autumn Session, Vijayawada, Uddaraju Raman, 1968.

Crow, B. 1995. *Sharing the Ganges*. New Delhi: Sage Publications.

Deb, P. 1985. Rural development and energy systems analysis. PhD Thesis, Indian Institute of Technology, Delhi.

Duggal, K.N. 1975. Three Level Optimization of an Agricultural Growth System. Ph.D. Thesis. Indian Institute of Technology, Delhi.

Eaton, D.J. (ed.). 1992. *The Ganges–Brahmaputra Basin*. Austin, TX: The University of Texas at Austin, LBJ School of Public Affairs.

Gupta, D.K. 1984. Conjunctive surface and groundwater development—Multilevel multiobjective analysis. PhD Thesis, Indian Institute of Technology, Delhi.

Hasan, Q. 2005. A Study of Sustainable Development of Waters of India, Ph.D. Thesis. Indian Institute of Technology. Delhi.

Khepar, S.D. 1980. System studies for microlevel irrigation water management. PhD Thesis, Indian Institute of Technology, Delhi, New Delhi.

Kothari, M. 2000. Sustainable Water Resources Development for a River Basin, Ph.D. Thesis, Indian Institute of Technology, Delhi.

Lieftnick, P., A.R. Sadov, and T.C. Creyke. 1968. *Water and Power Resources of West Pakistan: A Study in Sector Planning*. Baltimore: John Hopkins Press.

Malla, S.K., S.K. Shreshta, and M.M. Sainju. 2001. Nepal's water vision and the GBM basin framework, in *Ganges–Brahmaputra–Meghana Region—A Framework for Sustainable Development*, edited by Q.K. Ahmad, A.K. Biswas, R. Rangachari, and M.M. Sainju (eds.). Dhaka: The University Press Ltd.

Michel, A.A. 1967. The Indus Rivers. New Haven. Yale University Press.

Nehru, J. 1947. *The Discovery of India*. London: Penguin Books.

Panandikar, V.A.P. 1992. Foreword, in *Harnessing the Eastern Himalayan Rivers*, edited by B.G. Verghese and R.R. Iyer (eds.). New Delhi: Konarak Publishers.

Pfister, C. and P. Messerli. 1990. Switzerland, in *The Earth as Transformed by Human Action*, edited by Turner II et al. Cambridge: Cambridge University Press.

Prasad, R.K. 1981. Finite element simulation of groundwater system. PhD Thesis, Indian Institute of Technology, Delhi, New Delhi.

Rangachari, R. 2006. Bhakra-Nangal Project-Socioeconomic and environmental impacts. New Delhi. Oxford University Press.

Report on Land and Water in the Indus Basin. White House Department of Interior Panel on Waterlogging and Salinity in West Pakistan. The White House. Washington D.C. 1964.

Rothmans, J. and B. de Vries (eds.). 1997. *Perspective on Global Change—The TARGETS Approach*. Cambridge: Cambridge University Press.

Sarma, E.A.S. 1985. Energy policy of India. PhD Thesis, Indian Institute of Technology, Delhi.

Schwarz, H.E., J. Emel, W.J. Dickens, P. Rogers, and J. Thompson. 1990. Water quality and flows, in *The Earth as Transformed by Human Action*, edited by B.L. Turner II, W.C. Clark, R.W. Kates, J.F. Richards, J.T. Mathews, and W.B. Meyer. Cambridge: Cambridge University Press.

Singh, R. 1980. Study of Some Aspects of Water Resources and Agricultural Policy of India. Ph.D. Thesis. Indian Institute of Technology, Delhi.

Stone, I. 1984. *Canal Irrigation in British India*. Cambridge: Cambridge University Press.

Thangraj, C. 1987. Integrated water—Power systems planning. PhD Thesis, Indian Institute of Technology, Delhi.

Thapa, B. and B.B. Pradhan. 1995. *Water Resources Development—Nepalese Perspective*. New Delhi: Konarak Publishers.

Verghese, B.G. and R.R. Iyer (eds.). 1992. *Harnessing the Eastern Himalayan Rivers*. New Delhi: Konarak Publishers.

Wilson, H.M. 1903. Irrigation in India. US Geological Survey Water Supply and Irrigation Paper No. 87. US Department of Interior, Washington, DC.

Chapter 7

Bower, B.T. and M.N. Hufschmidt. 1984. A conceptual framework for analysis of water resources management in Asia. Natural Resources Forum, 8(4), 343–356.

Briscoe, J. and R.P.S. Malik. 2006. India's Water Economy-Bracing for a turbulent Future. New Delhi. Oxford University Press.

Chaturvedi, M.C. 2009b. *India's Waters—Development* (in press).

Chaturvedi, M.C. 2009c. *Revolutionizing the Development and Management of India's Waters* (in press).

Chaturvedi, M.C. 2011a. *Environmental Systems Management for Sustainable Development* (in press).

Chaturvedi, M.C. 2011c. India's Waters-Advances in Development and Management. Boca Raton, FL: Taylor & Francis.

Chaturvedi, M.C. 2011d. Ganga-Brahmaputra-MeghanaWaters-Development and Management. Boca Raton, FL: Taylor and Francis.

Chaturvedi, M.C. and Chaturvedi V. 2012. India-Tryst with Destiny (in press).

Frederiksen, H.D., J. Berkoff, and W. Barber. 1993. Water resources management in Asia, Vol. 1, Main report, World Bank Technical Paper, No. 212. World Bank, Washington, DC.

Gulhati, N.D. 1972. *Development of Inter-State Rivers—Law and Practice in India*. New Delhi: Allied Publishers.

Iyer, R.R. 2008. Water government, politics, policy, in *Governance of Water*, edited by V. Ballabh. New Delhi: Sage Publications.

Landes, D.S. 1969. *The Unbound Prometheus*. Cambridge: Cambridge University Press.

Landes, D.S. 1998. *The Wealth and Poverty of Nations—Why some are rich and some so poor*. New York: W.W. Norton.

Landes, D.S. 2000. Culture makes almost all the difference, in *Culture Matters*, edited by L.E. Harrison and S.P. Huntington. Basic Books: New York.

National Commission for Integrated Water Resources Development (NCIWRD). 1999. Report, Vol. I, Integrated Water Resources Development—A Plan for Action. Ministry of Water Resources, Government of India, New Delhi.

North, D.C. 1990. Institutions, *Institutional Change and Economic Performance*. Cambridge: Cambridge University Press.

Rothmans, J. and B. de Vries. 1997 (ed.), Perspective on Global Change-The TARGETS Approach, Cambridge, Cambridge University Press.

Chapter 8

Chaturvedi, M.C. 2007. Management of Ganga–Brahmaputra–Meghana Waters—People Oriented Global Leadership Project. Ministry of Water Resources, Government of India.

Chaturvedi, M.C. 2011a. *Societal Environmental Systems Management* (in press).

Chaturvedi, M.C. 2011b. *India's Waters—Environment, Economy and Development*. Boca Raton, FL: Taylor & Francis.

Chaturvedi, M.C. 2011c. *India's Waters—Advances in Development and Management*. Boca Raton, FL: Taylor & Francis.

Chaturvedi, M.C. 2011d. *Ganga–Brahmaputra–Meghana Waters—Development and Management*. Boca Raton, FL: Taylor & Francis.

Sen, A. 1989. Indian development: Lessons and non-lessons. *Daedalus*, 1184, (Fall), 369–392.

Appendix 2

Chaturvedi, M.C. 1976. *Water—Second India Studies*. New Delhi: Macmillan.

Chaturvedi, M.C. and P. Rogers (eds.). 1985. *Water Resources Systems Planning—Some Case Studies for India*. Bangalore: Indian Academy of Sciences.

Landes, D.S. 1969. *The Unbound Prometheus*. Cambridge: Cambridge University Press.

Landes, D.S. 1998. *The Wealth and Poverty of Nations—Why some are rich and some so poor*. New York: W.W. Norton.

Landes, D.S. 2000. Culture makes almost all the difference, in *Culture Matters*, edited by L.E. Harrison and S.P. Huntington. New York: Basic Books.

National Commission for Integrated Water Resources Development (NCIWRD). 1999. Report, Vol. I, Integrated Water Resources Development—A Plan for Action. Ministry of Water Resources, Government of India, New Delhi.

Rogers, P. 1967. A Systems Analysis of the Lower Ganges–Brahmaputra Basin. UNESCO, International Symposium on Floods and Their Composition, Leningrad, August.

Swaminathan, M.S., M.C. Saxena, M.C. Chaturvedi, M. Gopal, and O. Thanvi. 2002. *Lava ka Baas*. New Delhi: Centre for Science and Environment.

Appendix 1

Estimate of Water Demand by NCIWRDP 1999

According to the available estimates, the total withdrawal/utilization for all uses in the year 1990 was 553 km³ or 655 m³/person/year. Out of the total water utilized in the country, irrigation accounted for nearly 83%, followed by drinking and municipal use (4.5%), energy development (3.5%), and industries (3%). Other activities claimed about 6% of the total use.

Future demand will depend on population, economic development, and techno-economic–environmental and management policies. NCIWRD (1999) has worked out the long-term demands. It is considered in terms of the following: (1) irrigation, (2) domestic use, (3) industry, (4) power development, (5) inland navigation, (6) environment and ecology, and (7) compensation for evaporation losses from reservoirs.

Water Requirement for Irrigation

The water requirement is based on several policy and technological assumptions. It is considered that food self-sufficiency and, to some extent, export of food and non-food agriculture produce are essential for the country from both strategic and socioeconomic considerations. The requirement of food production would therefore mainly depend on the country's population and per capita income, and changes in dietary habits.

Land Use: Estimates have to be made of the likely future land use, yield of irrigated and unirrigated land, and water requirement in irrigated land for estimates of water demand for agriculture. All are interrelated.

The pattern of land use in 1993–1994 is given in Table A1.1. Out of the total geographical area of 328.726 million hectares (Mha), the reporting area was only 300.358 Mha. The forests and permanent pastures covered 68.4 and 11.3 Mha or 23% and 3.8%, respectively. Other non-agricultural uses accounted for 47.769 Mha or 16% of the land area. The total cultivable area was 184.256 Mha or 62% of the land area. This is 12.7% of the world's cultivable area, which is at 1475.09 Mha.

The total net sown area of the country has gradually increased from 118.75 Mha in 1950–1951 to 140.27 Mha in 1970–1971. Since then, the net sown area has been fluctuating between 140 and 142.5 Mha. The gross cropped area has increased from 131.89 Mha in 1950–1951 to 186.42 Mha in 1990–1991. The

TABLE A1.1

Land Use Pattern for the Year 1993–1994[a]

Sl. No.	Particulars	Area
1.	Total geographical area	328,726
2.	Reporting area	300,358
3.	Forest	68,421
4.	Area not available for cultivation	
	a. Area put to nonagricultural uses	22,035
	b. Barren and unculturable	18,975
5.	Other uncultivated land excluding fallow lands	
	a. Permanent pasture and other grazing lands	11,176
	b. Land under miscellaneous tree crops and grooves not included in net sown area	3657
	c. Culturable wasteland	14,468
6.	Fallow lands	
	a. Fallow lands other than current fallows	9703
	b. Current fallows	14,333
7.	Net sown area	142,095

Source: National Commission for Integrated Water Resources Development (NCIWRD), Report, Vol. I, *Integrated Water Resources Development—A Plan for Action,* Ministry of Water Resources, Government of India, New Delhi, 1999.

[a] In 1000 ha.

proportion of the gross sown area to net sown area or cropping intensity in that period was 130%.

Increase in gross cropped area is mainly because of expansion of irrigation. The gross irrigated area has increased from 22.6 Mha in 1950–1951 to 68.4 Mha in 1993–1994. The area irrigated from surface water was 30.87 Mha or 45% of the total irrigated area. Thus, out of the increase of 54.53 Mha in gross cropped area, nearly 84% or 45.8 Mha could have been due to the additional irrigation facility.

The ultimate irrigation potential of the country is estimated at 140 Mha, out of which 75.9 Mha would be from surface water and 64.1 Mha from a groundwater source. At that level, irrigation from surface water and groundwater would be 54.3% and 45.7% of the total irrigated area.

Cropping Pattern: The cropping pattern in the country has gone through several changes. Among food grains, from 1950–1951 to 1990–1993, the area under wheat has increased by 6%, whereas areas under jowar, barley, and gram have decreased. Among the non-food crops, areas under sugarcane, fruits and vegetables, and oilseeds have increased by 1%, 2%, and 6%, respectively, over the same period. However, the ratio of area under food grains and nonfood grains has remained more or less static at 2:1. The area under food grains was 67% of the cultivated area of 186.45 Mha in 1993–1994.

Out of the total gross irrigated area of 68.4 Mha in 1993–1994, 48.247 Mha or 70% was under food grain crops. This 70% is made of areas under rice, wheat, other cereals, and pulses at 30%, 31%, 5%, and 4%, respectively. An average of 79% of the rice-cropped area is irrigated in the country as a whole. However, the percentage of irrigated area to cropped area under rice varies from 14% to 100% among states, with the highest percentage being nearly 100% and reported in Haryana, Andhra Pradesh, and Tamil Nadu. Most of the total cropped area (81%–84%) under wheat and barley is irrigated. Bihar (20%), Karnataka (35.7%), and Tamil Nadu (13%) have less than 50% of wheat under irrigation. The average total rainfed (unirrigated) area in 1993–1994 was 118.18 Mha, out of which the area under food grain crop was 77.317 Mha. The percentage of food grain area to total rainfed cropped area was 66%.

Land Use Projections: Taking into account the available land and water resources and the trends in cropping pattern, the NCIWRD (1999) made the assumptions shown in Table A1.2 to estimate water requirements for irrigation.

Current Food Grain Yield and Yield Potential: State-wise land use, irrigation, and food grain production figures are available for 1993–1994 and have been given. Yields of various crops significantly differ from state to state. The yields of crops at a specific location depend not only on input management, cultural practices, and seed varieties but also on rainfall and irrigation.

The national-level average yields in the year 1991–1992 for all food grains under rainfed and irrigated conditions were 1.0 and 2.33 ton/ha, respectively, with an overall average of 2.062 ton/ha. However, the NCIWRD chose to adopt the figures given in Table A1.3 as those for which "good probability exists for achieving."

TABLE A1.2

Land Use Projections for Estimation of Food Production under Two Scenarios (High Demand Scenario and Low Demand Scenario)

Sl. No.	Particulars	2010	2025	2050
1.	Net area sown (Mha)	143.0	144.0	145.0
2.	Cropping intensity (%)	135	140–142	150–160
3.	Percentage of irrigated to gross cropped area	40–41	45–48	52–64
4.	Percentage of irrigated food crop area to gross irrigated area	70	70	70
5.	Percentage of rainfed food crop area to gross rainfed cropped area	66	66	66
6.	Percentage of surface water irrigation to total irrigation	47	49–51	54.3

Source: National Commission for Integrated Water Resources Development (NCIWRD), Report, Vol. I, *Integrated Water Resources Development—A Plan for Action*, Ministry of Water Resources, Government of India, New Delhi, 1999.

Notes: Food crop area includes area under all food equivalents, that is, horticulture, etc. The range against several rows depicts the low and high values corresponding to low demand and high demand scenarios.

TABLE A1.3

Future Food Crop Yield Projections[a]

	2010	2025	2050
Rainfed food crop yields	1.1	1.25	1.5
Irrigated food crop yields	3.0	3.4	4.0

Source: National Commission for Integrated Water Resources Development (NCIWRD), Report, Vol. I, *Integrated Water Resources Development—A Plan for Action,* Ministry of Water Resources, Government of India, New Delhi, 1999.

[a] In tons per hectare.

Remarkably, it may be pointed out that the yield figure adopted for 2050 is almost the same as that already achieved in China.

Depth and Efficiency of Irrigation: The net irrigation requirement (NIR) varies with the crop, climatic factors, and effective rainfall during the crop growing period, etc. There are a number of theoretical approaches, which also take into account the climatological factors. The gross irrigation requirement (GIR) or "Delta," that is, depth of irrigation at canal head, is a function of NIR and efficiency of conveyance (GIR = NIR/Overall irrigation efficiency). Conveyance and application losses, which are field losses contributing to overall irrigation efficiency from source to field application, are expected to vary widely with the type of strata through which the canal system passes; the material used and the quality of work in the canal lining; preparation of the field; and the type of soil, stream size, and method of irrigation. Although no national-level assessments of overall irrigation efficiencies from surface water and groundwater are available, NCIWRD (1999) adopted the figures as given in Table A1.4.

NIRs for each block/district in the country have been estimated for the prevailing cropping pattern using a climatological approach by the concerned agricultural department/agricultural universities. The national weighted average value of NIR is estimated at 0.36 m. The national weighted average values of GIR would be as shown in Table 4.7, based on efficiencies assumed and given in Table A1.5.

TABLE A1.4

Projected Overall Irrigation Efficiencies

	2010	2025	2050
Surface water (%)	40	50	60
Groundwater (%)	70	72	75

Source: National Commission for Integrated Water Resources Development (NCIWRD), Report, Vol. I, *Integrated Water Resources Development—A Plan for Action,* Ministry of Water Resources, Government of India, New Delhi, 1999.

TABLE A1.5

Projected National Average Values of GIR (or "Delta")[a]

Source of Water	Estimated by Central Ground Water Board	Assumed Depth of Irrigation of "Delta"		
		2010	2025	2050
NIR	0.36			
Ground water: Assumed GIR		0.52	0.51	0.49
Surface water: Assumed GIR		0.91	0.73	0.61

[a] In meters.

Water Requirement for Irrigation: For meeting the domestic demand for food grain, the water demand for irrigation for various years, based on the assumptions made above, has been estimated as shown in Table A1.6. It is projected that by the year 2050, the water demand for irrigation will be around 628 km³ for low demand and 807 km³ for high demand.

Water Demand for Domestic Use

The water demand for domestic use depends on the level of service contemplated to be provided or, more appropriately, demanded by the people. In developing countries, it is usual that planning is carried out in terms of certain norms, some of which are not even observed.

Norms have been suggested by several agencies. The existing national average of water supply, at the point of supply for Class I cities and Class II cities, are 182 and 103 liters per capita per day (lpcd). The norms adopted for planning by the NCIWRD (1999) are given in Table A1.7.

The national water requirement for drinking and municipal uses at different points of time is estimated as given in Table 4.10.

It is assumed that 70% of the urban and 30% of the rural water requirement would be met by surface water resources, with the remaining percentage coming from groundwater sources. Thus, the total water requirement use for rural and urban areas by the year 2050 is estimated at 90 and 111 km³ for the two scenarios. The water requirement of more than 34 million cities for the year 2050 would be 13.2 km³, which is included in the above estimate.

Water Requirement for Industries

It is difficult to estimate the water requirement of the industries. The estimate of the NCIWRD (1999) is as follows. The estimated water requirements

TABLE A1.6

Water Requirement for Irrigation

Sl. No.	Particulars	Unit	2010 Low Demand	2010 High Demand	2025 Low Demand	2025 High Demand	2050 Low Demand	2050 High Demand
1.	Food grain demand	Million tons	245	247	308	320	420	494
2.	Net cultivable area	Million hectares	143	143	144	144	145	145
3.	Cropping intensity	%	135	135	140	142	150	160
4.	Percentage of irrigated to gross cropped area	%	40	41	45	48	52	63
5.	Total cropped area	Million tons	193	193	202	204	218	232
6.	Total irrigated cropped area	Million hectares	77	79	91	98	113	146
7.	Total unirrigated cropped area	Million hectares	116	114	111	106	105	86
8.	Food crop area as percentage of irrigated area	%	70	70	70	70	70	70
9.	Food crop area as percentage of unirrigated area	%	66	66	66	66	66	66
10.	Food crop area—Irrigated	Million hectares	54.1	55.4	63.5	68.7	79.2	102.3
11.	Food crop area—Unirrigated	Million hectares	76.4	75.2	73.5	70.2	69.3	56.7

No.	Item	Unit						
12.	Average yield—Irrigated food crop	Ton per hectare	3	3	3.4	3.4	4	4
13.	Average yield—Unirrigated food crop	Ton per hectare	1.1	1.1	1.25	1.25	1.5	1.5
14.	Food grain production from irrigated area	Million tons	162	166	216	216	317	409
15.	Food grain production from unirrigated area	Million tons	84	83	91	91	105	85
16.	Total surrogate food production	Million tons	246	249	307	307	422	494
17.	Assumed percentage of potential from surface water to total irrigation potential	%	47	47	49	49	54.3	54.3
18.	Irrigated area from surface water	Million hectares	36.3	37.2	44.5	44.5	61.4	75.9
19.	Irrigated area from groundwater	Million hectares	40.9	41.9	46.3	46.3	51.7	70.3
20.	Assumed "Delta" for surface water	Meters	0.91	0.91	0.73	0.73	0.61	0.61
21.	Assumed "Delta" for groundwater	Meter	0.52	0.52	0.51	0.51	0.49	0.49
22.	Surface water required for irrigation	Cubic kilometer	330	339	325	366	375	463
23.	Ground water required for irrigation	Cubic kilometer	213	218	236	245	253	344
24.	Total water required for irrigation	Cubic kilometer	543	557	561	611	628	807

Source: National Commission for Integrated Water Resources Development (NCIWRD), Report, Vol. I, *Integrated Water Resources Development—A Plan for Action*, Ministry of Water Resources, Government of India, New Delhi, 1999.

TABLE A1.7

Norms for Domestic Water Supply at Different Points of Time[a]

Population Type	Year 2010	Year 2025	Year 2050
Class I cities	220	220	220
Other than class I cities	150	165	220
Rural	55	70	150

Source: National Commission for Integrated Water Resources Development (NCIWRD), Report, Vol. I, *Integrated Water Resources Development—A Plan for Action*, Ministry of Water Resources, Government of India, New Delhi, 1999.

[a] In liters per capita per day.

for industrial development are 37, 67, and 81–103 km^3 for the years 2010, 2025, and 2050, respectively. The requirement of 103 km^3 in the year 2050 corresponds to the present rate of use of water, whereas the requirement of 81 km^3 in the year 2050 assumes that there will have been a significant breakthrough in the adoption of water-saving technologies for industrial production. The latter figures have been used in the estimation of total water requirement for the year 2050. Of the total water requirement, 70% is estimated to be met using surface water sources and the remaining 30% using groundwater sources.

Water Requirement for Power Development

The power industry requires substantial amounts of water and is generally considered separately. Hydro stations do not use any water. Considering all factors, water requirement for the energy/power sector has been estimated for low- and high-demand scenarios at 18 and 19 km^3, 31 and 33 km^3, and 63 and 70 km^3 for the years 2010, 2025, and 2050, respectively. Eighty percent of the water requirement is expected to be met using surface water sources and the remaining 20% using groundwater sources.

Water Requirement for Inland Navigation Development

It is not considered that any special provision of water supplies will be required for meeting the navigation requirements. However, a small quantity has been provided if the need arises. The estimates are 7, 10, and 15 km^3 for the years 2010, 2025, and 2050.

Water Required for Environment and Ecology

There has been increasing concern for the conservation of the environment and its restoration. Sustainable management of the environment is currently one of the most important issues. Action on several fronts will be required, as discussed in Chapter 14. Demand of water in this context may be considered on two counts: (1) requirement of water for the restoration of forests and grasslands, if any, and (2) demand for water to restore the deprived, once-free-flowing rivers.

Afforestation is a top priority in India. Its dense forest cover, which was estimated to be about 40% in the last century, has now been reduced to less than 14%. Afforestation can best be done by utilizing precipitation and soil cover, and no special provision of water is required for this purpose.

There are also urgent needs for the abatement of pollution and for managing water quality. Large stretches of rivers have been converted into open sewers. There is urgent need to manage return flows and to maintain certain minimum flows to preserve water quality. The NCIWRD (1999) considered that the quantity of water required for dilution and treatment of sewage in all the rivers in the country is just not available. Alternative methods will have to be evolved and operationalized. A token provision of 5, 10, and 20 km³ of water for this purpose for the years 2010, 2025, and 2050, respectively, has been made.

Water Requirement to Compensate Evaporation Losses from Reservoirs

Compensation for evaporation losses from reservoirs has to be made. There have been several ways of estimating the evaporation requirements. Evaporation losses from reservoirs have been estimated to be 42, 50, and 76 km³ for the years 2010, 2025, and 2050, respectively.

Total Water Requirement

The total water requirement for various sectors of use is summarized in Table A1.8. According to NCIWRD (1999) estimates, total water requirements of the country would be 694–710, 784–843, and 973–1180 km³ by the years 2010, 2025, and 2050, respectively, depending on the low-demand and high-demand scenarios. Irrigation would continue to have the highest water requirement. By the year 2050, these will be 628–807 km³ for irrigation (68%), followed by domestic water use, including drinking and bovine needs, at

TABLE A1.8

Water Requirement for Different Uses[a]

Sl. No.	Uses	Year 1997–1998	2010 Low	2010 High	%	2025 Low	2025 High	%	2050 Low	2050 High	%
	Surface Water										
1.	Irrigation	318	330	339	48	325	366	43	375	463	39
2.	Domestic	17	23	24	3	30	36	5	48	65	6
3.	Industries	21	26	26	4	47	47	6	57	57	5
4.	Power	7	14	15	2	25	26	3	50	56	5
5.	Inland navigation		7	7	1	10	10	1	15	15	1
6.	Flood control		–	–	0	–	–	0	–	–	0
7.	Environment (1) afforestation		–	–	0	–	–	0	–	–	0
8.	Environment (2) ecology		5	5	1	10	10	1	20	20	2
9.	Evaporation losses	36	42	42	6	50	50	6	76	76	6
10.	Total	399	447	458	65	497	545	65	641	752	64
	Groundwater										
1.	Irrigation	206	213	218	31	236	245	29	253	344	29
2.	Domestic and municipal	13	19	19	2	25	26	3	42	46	4

3.	Industries	9	11	11	1	20	20	2	24	24	2
4.	Power	2	4	4	1	6	7	1	13	14	1
	Total	230	247	252	35	287	298	35	332	428	36
	Grand total	629	694	710	100	784	843	100	973	1180	100
Total Water Use											
1.	Irrigation	524	543	557	78	561	611	72	628	807	68
2.	Domestic	30	42	43	6	55	62	7	90	111	9
3.	Industries	30	37	37	5	67	67	8	81	81	7
4.	Power	9	18	19	3	31	33	4	63	70	6
5.	Inland navigation	0	7	7	1	10	10	1	15	15	1
6.	Flood control	0	0	0	0	0	0	0	0	0	0
7.	Environment (1) afforestation	0	0	0	0	0	0	0	0	0	0
8.	Environment (2) ecology	0	5	5	1	10	10	1	20	20	2
9.	Evaporation losses	36	42	42	6	50	50	6	76	76	7
	Total	629	694	710	100	784	843	100	973	1180	100

Source: National Commission for Integrated Water Resources Development (NCIWRD), Report, Vol. I, *Integrated Water Resources Development—A Plan for Action*, Ministry of Water Resources, Government of India, New Delhi, 1999.

[a] Quantity in cubic kilometers.

about 90–111 km³ (9%), industries at 81 km³ (7%), evaporation at 76 km³ (7%), power generation at 63–70 km³ (7%), and environmental need, navigation needs, etc. at 35 km³ (3%). (Percentages are in terms of total requirements.)

Basin and State Water Requirement

There are vast differences in the availability of water from one basin to another and from one state to another. As the basic unit of an integrating water plan is a basin or state, it is important to analyze the water balance between availability and requirement at a basin/state level. This has been carried out by the NCIWRD (1999) for low- and high-demand scenarios. The overarching objective of the exercise was achievement of food self-sufficiency at the national level.

Return Flows

The use of water for different purposes is partly consumptive and partly nonconsumptive. The latter is returned to the surface water and groundwater components or to the hydrological cycle albeit in a polluted state. It will be necessary to manage it in terms of its quality. It can also be reused after proper treatment or even without it.

Water Balance

Attempts have been made to develop the likely picture of the availability and requirement of water up to the year 2050. The result is just an approximate estimate. Summaries at the three levels, namely, national, basin, and state levels, were developed. The three exercises give approximately the same results.

Conclusion

According to the National Commission on Integrated Water Resources Development Plan, the central message is that by 2050, the country's requirements would barely match the estimated utilizable water resources. Therefore, according to the official policy, interlinking of the rivers has to be undertaken immediately.

Appendix 2

Political Economy of Water

In this study, we have examined development and management of waters from a scientific and engineering perspective. We have also given the governmental management perspective. However, water has to be used, and even managed, by the society. How does the society perceive water?

We have considered that the best approach to understand this issue is to present some of the experiences of the author as he grew up and, later, as he became closely associated with the development and management of water and has continued using water. It becomes a bit personal, but perhaps it may illuminate the subject best and with authenticity.

The author's earliest perception of water is as a child, as he grew up in a small town, Etah, in Uttar Pradesh (UP), where his father was posted as the Civil Surgeon with the Government of UP. The story relates to the early 1930s. The population of then undivided India was still about 300 million. The drinking water was obtained from wells, drawn up by human agency. There was no arrangement for public water supply, which continues in the small towns, much less the villages. Arrangements for serving water to people during their everyday work in hot summer months were made by philanthropic people through the establishment of "pyoos," which consisted of a man who dispensed water to passersby. There were different arrangements for Hindus and Muslims. For the latter, a big earthen water pot with a drinking mug was placed under a shade. One can recall that the edicts of Ashoka contain reference to these activities in ancient India. Apparently, India had not changed much. Another incident that is in remembrance is the big fair on the banks of Ganga at Soron, which was considered a holy town. The river Ganga had dwindled to a small stream in a large river course, mostly sand.

We were transferred to Gorakhpur shortly thereafter. The water environment was entirely different. One could feel the distinct difference between the arid west and the wet Terai. There were many trees, amounting to almost a woodland, in the compound. In Etah, the well from which water was fetched by the domestic servant was about a hundred feet below the ground. Here one could touch the water in the well. We were told that the water was not potable and that we had to take in boiled water. A vivid recollection of a scene of Rapti River, only a tributary of the mighty Ghagra, still haunts me. There was the annual flood. There was one motorboat in charge of the collector, who was an Englishman. We took a ride in the motorboat with him. Heads of people trying to avoid drowning could be seen. Some people were still perched on the trees, now engulfed by the river.

Similar scenes of the river in Bahraich, another Terai city to which our father was transferred later, although not so frightening, are also vivid. We had a big bungalow in the civil lines, where few senior officials lived. There was a big plot of land, maybe about 100 acres, on which it was located. However, the construction, dating back to long past, was primitive and came with a thatched roof. Water supply was from a hand pump, and the sanitary arrangements were still primitive. There was no electricity. We had to get by with lanterns. There was a small Government school providing education up to high school. The drinking water in the school was not considered safe, so we carried our own drinking water in bottles or thermal containers. This is still the life of a vast majority of the people in India, living in cluttered cities or villages, where life continues to be even more primitive.

Our father retired in 1940, and we moved to Mathura, in the west, on the banks of River Yamuna. There were several distinct differences. One remarkable difference was again brought out as to how the water environment shaped life in general. Mathura, like Etah, was in the arid west, but being a bigger and slightly developed city, its harshness as an arid city was not as marked. Perhaps being located along a major river, Yamuna, was also a factor. Mathura is a holy city, the land of Lord Krishna. People would go to the river to have a dip in the morning. Many got water from Yamuna for drinking, although the river was badly polluted.

The development of water in British India, when the author got engaged in the sector, was undertaken by officials of the Irrigation Department of the Government. Providing water supply for domestic and sanitary purposes was the most important activity, but it was given low priority historically and in British India (a situation that continues even today). The officers were the graduates from the then Thomson College of Civil Engineering, Roorkee, which was established in 1846 for this purpose. They were appointed as Apprentice Engineers in the Irrigation Department, and after about a year or so, they were confirmed as Assistant Engineers in the Department.[1]

Upon passing from Thomson College of Civil Engineering, Roorkee, in 1946, I was appointed as Apprentice Engineer in His Majesty's Government in the Sarda Power House Division. It had been established for undertaking the construction of the Sarda Power House, a run-of-canal power house. The arrangement was that the planning and design were done by the departmental engineers while a contractor carried out the construction. This was how the planning and design of the engineering works were undertaken in the army, and the practice was introduced in the public works department because all activities in the earlier phases of the British rule were carried out by Army officers. The contractors were small entities who hardly had any engineering training or knowledge. Engineering activity in India was limited. Paradoxically, the scene remains essentially the same today, except for some changes very recently when a few large engineering construction firms have started to emerge. Even today, the scientific capability of even the major firms is very limited.

Departmental engineers did not have much technical capability beyond having passed from Thomson College a long time back, which was essentially an officer's training school imparting departmental practice without much scientific engineering capability. Research was totally unknown. The job of the engineering officer in the department was to administer and supervise the running and managing of canals. The Sarda Power House was one of the first divisions that had been established to undertake some engineering activity in the latter period of the British administration.

There were only two young engineers (including me) who were engaged in the design, the other one being one or two years my senior. There was no senior engineer to guide us. I completed the design as best as I could, my technical capability being almost negligible. My only qualification was that I was one of the bright students in the class, which was the reason that I was tapped for planning and design. I submitted the design to the Executive Engineer, and, to my horror, the orders were immediately given to begin construction, which involved quite a major structure. There was even no checking done.

I completed my apprenticeship period in a few months because there was a shortage of engineers. I was appointed as Assistant Engineer in the department and was posted in the Design Division of Nayar Dam. Nayar Dam was a very major structure based on the proposal, one of the world's biggest dams at that time. A full-fledged division headed by an Executive Engineer and staffed with four Assistant Engineers, which was the departmental regular organizational system, had been established for the design. This was the first such division in the department tapped exclusively for design. The two senior officers, the Superintending Engineer and the Executive Engineer, who had a reputation in the department of being bright in theoretical engineering, headed the organization. They had been sent to the USA for 6 months of training at the US Bureau of Reclamation, which was considered the top engineering organization for planning and design of dams at that time.

I was assigned the task of working on the backwater so that the power house location could be fixed. I did not have a ghost of an idea as to how it was to be calculated, nor did anybody else in the organization. Fortunately, engineering books had been obtained for the library from the USA, as part of the officers' US training. I could get guidance from one.[2] I moved on to more complex designs, but the construction of Nayar Dam was given up because the foundations were considered unsuitable and the Division was wound up. I was posted back at the Sarda Power House, apparently because I was considered to have had obtained some experience in design. I, along with my other young colleagues, designed the structure, which was major engineering work; it was constructed based on our designs, without any checking from the senior engineers nor any detailed drawings. To my surprise, the structure still stands.[3]

During the later period, I was posted in the departmental construction of a portion of the Sarda Power Channel. I stayed in a tent, right on the site of the construction. Two experiences are memorable. One, I had an opportunity to meet some local villagers. They paid me considerable attention and respect because I was an irrigation department officer, expecting that I could provide some canal water to them. They were very disappointed to learn that I was only concerned with the construction of a power channel. The second was discovering that the lower departmental staff was totally corrupt. The officers, until then, were, however, scrupulously honest.

There was a demand from Himachal Pradesh for engineers because the engineering organization had not yet been set up in the newly constituted state. That gave me a valuable experience in the problems and challenges of water in the Himalayas, beyond the technical orientation of constructing dams, which totally usurps the engineering mind. The people face an extreme shortage of water for domestic needs. The only source is some springs at some locations in the hillside or some small hill streams. They are used for providing irrigation through the construction of long contour channels. Their need for water was expressed to me by an important resident of one major village, who promised to trade all his fortune for the construction of any means of providing water for the village. The scene continued, as I have noted, when I was working on the Board of Consultants of the hydroelectric projects of Ganga basin until recently.

I came back to UP and was assigned to work on the design of the Rihand Dam, located on a tributary of the river Son in the Vindhyas in southeast UP. It was a major structure based on international standards. I had been elevated to the rank of Executive Engineer and was the second in command. By this time, thanks to the advice of the Bureau of Reclamation, which was guiding the design of the Bhakra Dam, separate design organizations in line with US practice started to be established in India. It was a big improvement. We also had some US advisors. However, we found that they were not highly qualified educationally and only had some real-life experiences. It was these experiences that prompted me to go to the USA upon completion of the Rihand assignment. In 1959, under a Fulbright scholarship, we, my wife and I, went to the USA on a boat. We proceeded to the University of Iowa, one of the world's leading universities in the studies of water.[4] We had the opportunity of visiting Paris and New York on the way. We were in a different world, and we hoped to contribute to our country's transformation upon our return.

The Indian Institutes of Technology (IIT) was being established with the support of the friendly foreign countries and their engineering institutions—IIT Bombay with the Russians, IIT Madras with the Germans, and IIT Kanpur with a consortium of nine US universities headed by the Massachusetts Institute of Technology, along with Princeton University, University of Berkeley, University of Illinois, California Institute of Technology, Purdue University, Ohio University, University of Michigan, and the Carnegie Institute of Technology. Although the overall framework was similar, there was considerable difference between each unit. IIT Delhi, which was first established as the Delhi College of Engineering at a new location, had a considerably old faculty. It was, however, also raised to an IIT standard as IIT Delhi.

Upon my return in 1962, I was offered the assignment as Professor and Founder Head of the Department of Civil Engineering at IIT Kanpur; I was to become the oldest and most senior faculty member at 37 years, as I had gone to the USA after having worked in India for about 10 years in the profession (which was turning out to be a great advantage). Vipula joined the Humanities Department. We were all very young, but the US faculty, who were much senior and experienced, decided that the Indians should be heading the departments, with them being essentially responsible for establishing the institute. They were, however, very active collaborators and were essentially the founders of the institute. We all lived together at the campus housing, which was being constructed at an increasing pace.

Besides the rich academic experience, there were two illuminating experiences. According to the U.S. principle that the university has to serve the society, one of our U.S. colleagues suggested that we should go to the villages around the campus to find out what they expected of us.[5] He had devised an interesting technique of conducting this experiment. As we reached a village and got out of his car, children would flock around. He would focus his Polaroid camera on them, chant "jadu, jadu" (magic, magic), give their picture to them, and had rapport established. They would take us to the elders. He would repeat the trick and again have rapport established. We would start interaction with the villagers. On his initiation, the question was what is their first priority that we should address? The answer, unanimously and unhesitatingly, was "water." He would follow up with "for drinking purposes?" The answer would be "No, no! The girls fetch the water from the village pond. We want water for land, for irrigation." The assembly would giggle openly to learn that a woman was a professor at a male university. We visited several villages around the IIT Kanpur campus, and the response was always the same.

The other interesting experience was that for the U.S. faculty, the first priority was water for drinking purposes and sanitation. A senior professor from the University of California, Berkeley, was invited to IIT Kanpur to guide us in establishing the area. We went around the country and found that it was being taught only at three universities in India, one being the Medical College in Calcutta, which collaborated with Harvard University. We laid particular emphasis on it. Now, IIT Kanpur has become one of the leading institutions in the country in the area of public health engineering, which is still neglected in India.[6]

The Government of UP had undertaken the construction of the Ramganga Project. I rejoined the Government Irrigation Department and was put in charge of the design of a major engineering structure, the Ramganga Dam. It was the first major dam in the Himalayas in the Ganga basin. We designed it entirely on our own. A Board of Consultants consisting of some of the most eminent engineers in the country and the United States, with Dr. A.N. Khosla and Dr. A.C. Mitra as Chairman and Vice-Chairman, respectively, had been consulted to review and approve the designs for undertaking the construction of the major water resources in the Himalayas. Under the arrangement, the Director of Designs for the Beas Dam in the Indus basin participated in the review of the Ramganga Dam.

As Member of the Board of Consultants, I had the privilege of working with some of the leading Indian engineers, such as Dr. A.N. Khosla and Dr. A.C. Mitra, who paradoxically my gurus and under whom I had started my career. It was an inspiration to work under the guidance of these great Indian engineers and a pleasure to learn that they appreciated my work. They were dedicated people. Even in the profession, generally, the professional standards were high; there was commitment, and there was complete honesty. It was still the Nehruvian India.

Upon completion of the design of the Ramganga Project, I realized that I had much to learn in the field of water resources, in which a new chapter was being written, particularly by the Harvard scientists. My wife, Vipula, who had been working at the Lucknow University, and I had been invited to join IIT Delhi. I did so as the Head of the Department of Applied Mechanics. Vipula was promised an appointment in the Humanities Department. Until then, she joined the US Educational Foundation. Unfortunately, this promise was never kept, so she joined the neighboring Kurukshetra University. Despite all our personal difficulties, it was, in a way, a better deal, as she was working in her own field of specialization, where she soon achieved the highest levels.

IIT Delhi was being established with the support of some British Universities, particularly the Imperial College in London. It was a big disappointment, particularly in the context of our U.S. and IIT Kanpur experience.

Some fundamental changes started taking place. One, there was gradual deterioration in the culture of the profession in India. Corruption among officers, under the patronage of the political masters, increased rapidly. Conjunctively, the professional commitment and capability deteriorated equally rapidly.

On the other hand, advances were taking place in the field of water resources engineering. Besides the advances in the science and art of engineering, coupling was taking place with economics. The increasing advances in systems analysis were almost transforming engineering. This led to an interesting chance experience, resulting in a valuable opportunity to work closely with Dr. K.L. Rao, then Minister of Irrigation in the Government of India. One young faculty member at Harvard University, Peter Rogers, had written a paper applying the modern emerging science of systems planning to the Ganga basin development, focusing on activities in East Pakistan, to which Harvard University professors served as advisers (Rogers 1967). It was presented at an international symposium held in Leningrad. It was a simple academic study that only demonstrated the application of the new tool of systems analysis. It came to the attention of the Indian Government officers and was interpreted as an attempt on the part of U.S. policy advisers to prejudice the development of Ganga, which was disputed by the two countries but becoming in favor of East Pakistan, an ally of the United States. The officials were ignorant of the modern advances, and I was requested to explain it to the Honorable Minister. Dr. K.L. Rao appreciated my presentation, and a close rapport developed. He was appreciative of the revolutionary advances taking place in the field of water resources and extended full encouragement and support to my efforts to contribute to the advancement of the profession in India. However, the profession was totally indifferent, and we could not achieve anything.

A striking contrast was provided by my association with Harvard scientists, which started taking place through a number of scientific studies with which we were associated. A scientific conference on India's water resources was organized in India, sponsored by the Indian Academy of Sciences and the U.S. Academy of Sciences. Discussions about the storage of the monsoon precipitation took place. The subject was studied by Roger Revelle and me, who attended the conference, first independently and later collectively, as discussed in Chapter 5. We came to the conclusion that the monsoon waters can be stored very effectively through schemes called the Ganga Water Machine by Revelle and the Swadeshi Ganga Water Machine by me, which had some basic differences. I discussed the idea with Dr. K.L. Rao, who appreciated it and issued orders for field testing. The proposal was vindicated. However, no further action could be taken because he left shortly afterward, and since then, the interlinking of India's rivers has become an obsession. Revelle, a great scientist, was very conscious of the challenges that India faced and continued his efforts in official and personal capacity, but the official agencies ignored them because they were obsessed with the idea that the USA was an ally of East Pakistan, with whom we had a dispute regarding sharing of Ganga waters. The Ford Foundation was actively engaged in supporting advances in water resources development because they considered that India would be facing tremendous challenges over time on account of increasing population and the demands of a growing economy. The foundation supported two activities. They asked some identified scientists to write a small 100-page book on six subjects, orienting the people on the futures challenges in these areas. These were called Second India Studies, as it was considered that by the turn of the century, a Second India would be born. I was working in the area of water and energy, and I was asked to write on these subjects. I was about to leave for Harvard, and I passed the energy book to a younger colleague of mine, writing only the *Water—Second India Studies* (Chaturvedi 1976). It was well publicized, with the release by then Government of India's Minister of Irrigation, Shri K.C. Pant, himself.

It was a great education for me. Instead of focusing only on projects, which I continued as a Member of the Board of Consultants of the Major Projects, I started working in the emerging field of water resources systems planning.

This was again supported by the Ford Foundation in two ways. They encouraged the training of some senior professionals and academics by supporting a 6-month study at Harvard University, focusing on some specific problems.[7] In addition, they encouraged the education of young professionals in India by taking up a project focusing on the development of the water resources of the state, starting with the Indus basin in Punjab. Two young research scholars worked with me, leading to undertaking a research study in the context of a doctoral degree. This provided me with the understanding of the attitude and capability of the profession. It was clear that the profession was no longer interested in engineering. I had the privilege of working closely with the Punjab design engineers in the past in the context of Bhakra and later Beas Dam. They were the most advanced group in India at that time. They had produced the most notable engineers of India, such as Dr. A.N. Khosla and Dr. Kanwar Sain, but here it was, totally bankrupt and indifferent to engineering.

A dramatic experience helped my understanding reach a climax. One of the research scholars had finished his doctoral work. He went on to get married. The marriage took place with the daughter of a Chief Engineer in Punjab. He, along with his bride, came to me for my blessings. Simultaneously, he declared that he would not be submitting his completed doctoral dissertation for the award of a degree because he was told by his father-in-law that if he got a PhD tag, he would be posted in planning and design, which had no "scope."

This was not an isolated experience. The Ford Foundation encouraged a project about training and education of a dozen officers from UP. The project envisaged education and research leading to a PhD. After the first phase, which led to the awarding of an MTech, nine of them declared that they would no longer continue because if they earned a PhD, the tag would confirm their posting in planning and design, which had no scope.

In contrast, the attitude of U.S. research scholars was impressively different. A young research scholar from Harvard, John Briscoe, who worked closely with me while I was there, came one evening to my house at the IIT Delhi campus. He had decided to spend a year at a Bangladesh village to study the energy cycle at the village level of economy. He asked me the next morning as to how one went to ease himself, in Indian style, if there were no toilet facilities in the village where he planned to work. I told him and declared that now he was an India expert.[8] As he told me later, he stayed in a village and had to use the water from the pond for ablutions in the morning after the defecations in the open and for drinking purposes.

One recent experience provided illumination and a close interaction with the rural community. A scheme had been proposed by Prof. G.D. Agrawala, a former colleague in the Irrigation Department and at IIT Kanpur, that water could be developed by rural communities themselves through the construction of small earthen bunds on the developing village streams. He worked with the village community and provided the basic design of the bund. Some were successfully constructed in central highlands on small streams joining Yamuna. The scheme was encouraged and supported by the Centre for Science and Technology, established by Anil Agrawala, a former student of G.D. Agrawala under the much publicized rainwater harvesting, to build small storage structures. A recent construction at Lava ka Baas received objections from the Government officials on the plea that all water belonged to the state and that the development can only be undertaken by the state. They went on to demolish a small bund that was recently constructed by the villagers. I was called by the state as an expert to give my opinion (Swaminathan et al. 2002). I visited the village and gained first-hand experience of the village conditions. It was pathetic to see the condition of the people and their desperate need for water, which were just as bad as what I had seen 50 years ago.

It, however, also provided confirmation of the NCIWRD (1999), Government of India Commission, that even these small activities should be undertaken under the guidance of engineers. On examination of the structures, it was shocking to see how unscientific the activity was, understandably because of the ignorance of the villagers, even if the structure was small. I was able to save it from the fury of the official agencies, but the struture was washed away in the first rains.

I had the privilege of working closely with the Government of India at the highest levels, with the support of the Government, for modernization, but hardly with any results. The officials were busy in their bureaucratic gyrations. The efforts continue. I have been aiming at a collaborative India–Nepal development because it is crucial for the development of India. I have developed a novel idea called the Chaturvedi Water Power Machine. It has received an encouraging response from the highest levels. The major developments will take place in Nepal. It occurred to me that implementation in Nepal will take time, so I asked why we should not start with some projects that can be undertaken in India immediately. I phoned one of my former students in UP, who holds the most senior position. His response was "Sir, the objective function here is making money for our political masters and us, not engineering. Have you not read today's papers about the murder of one of my young officers who refused to give his quota to the Chief Minister's birthday celebrations?"

Conclusion

The review shows that the society is not conscious of the possibilities of the development that humanity has achieved. This is not realized even in basic terms such as better living conditions, easy availability of potable water, proper sanitation facilities, transport, and, above all, basic education. The prejudices against the female members of the society are still those of savage conditions. Even in economic activities, people are not conscious that their well-being can be dramatically increased. Agriculture, on which the vast majority depends, can be transformed if science and technology is brought to bear on it, in which scientific use of water is crucial. They cannot be expected to even visualize the danger that the society faces if urgent rapid development of the society is not undertaken, which becomes more threatening with the ongoing climate change.

The managers of the society, political and bureaucratic, except for some honorable ones, are totally indifferent to the people's concerns, even with the historic precepts immortalized in religion and, recently, in the person of Mahatma Gandhi. They are totally corrupt, engrossed in seeking personal benefits.

Landes, an eminent economic historian, has made an exhaustive study of the development of the Western world, bringing out the important role that technology has played (Landes 1969). He followed the study to explore why some countries are so rich and some are so poor (Landes 1998). After a monumental study, he came to a simple conclusion: culture makes all the difference (Landes 2000).

The review of the state of affairs in India fully supports Landes's conclusions. Water resources in India are in urgent need of the latest technology, but the culture totally prohibits it. The people who need water have no say in the development, and their perceptions are of an underdeveloped society. The technological capability of the society is poor. The decision makers and the executive agency, the political masters, and the officers are totally corrupt and indifferent to development.

Notes

1. There used to be a Class I and Class II at the earlier stage, but by the time the author joined the service, Class I had been abolished. The author graduated from the college in 1946, which was the last batch of the Thomson College of Civil Engineering, after which it became Roorkee University.

2. The experience has interesting reference to the subject, bringing out the international aspect of the subject under discussion. The book was *Posey's Open Channel Flow*, the first engineering book on the subject. The technique of calculation was Bresse's method of calculation of backwater curves. Bresse was one of the early, perhaps the first, professors of engineering at the Ecole Polytechnique as it was being

established in 1890 in France. Posey was a professor at the University of Iowa, Iowa City, IA, which, I would soon discover, was the leading university in this area and to which I applied for higher studies later. The doctoral students at US universities in those days had to learn two foreign languages (only to learn a bit over a one-semester course), so that they may be able to study journals and papers by European scientists. My doctoral guide, Professor Hunter Rouse, one of the most eminent scientists of the time, used to give assignments to the doctoral students, in one course entitled History of Hydraulics, to study the contributions of two scientists. The choice of the scientists was made based on the foreign language that the student had learned. I learned French and Russian, and the work of two scientists, one from each country, became my subject of study. The reference is to give a perspective of development of science in the West and in India.

3. I had the privilege of being invited to visit it, on some important occasion, with a team of eminent foreign scientists. I was frightened to cross while inside the car the bridge I had designed, although it was still standing, and chose to get out and cross on foot.

4. The US experience brought out that even in the USA, the problem of lack of professional experience on the part of the faculty and lack of engineering educational competence on the part of the professionals exists. Nevertheless, they are conscious of it.

5. He considered that the introduction of a female would facilitate establishment of the rapport.

6. The experience in this context was very interesting. IIT Kanpur had invited one U.S. professor to advise us about setting the program. I met him at the airport, and as we drove back through the city, he saw lots of people roaming around. He asked if it was some national holiday that day. I had to tell him that it was an everyday scene.

7. The project was coordinated by Dr. Peter Rogers and me. The studies have been released in book form (Chaturvedi and Rogers 1985).

8. The prophecy turned out to be true. He became Chief Advisor to the World Bank, New Delhi, on water and environment and is now a professor at Harvard University.

Index

Page numbers followed by f and t indicate figures and tables, respectively.